P9-AQL-214

Titles in This Series

Titles in This Series

Titles in This Series

Titles in This Series

CONTEMPORARY MATHEMATICS

117

Continuum Theory and Dynamical Systems

Proceedings of the AMS-IMS-SIAM
Joint Summer Research Conference
held June 17–23, 1989,
with support from the National Science Foundation
and the Army Research Office

Morton Brown
Editor

American Mathematical Society
Providence, Rhode Island

The AMS-IMS-SIAM Joint Summer Research Conference in the Mathematical Sciences on Relationships between Continuum Theory and the Theory of Dynamical Systems was held at Humboldt State University, Arcata, California, on June 17–23, 1989, with support from the National Science Foundation, Grant DMS-8613199 and the Army Research Office, Grant DAAL03-89-G-0038.

1980 *Mathematics Subject Classification* (1985 *Revision*). Primary 54F20, 54F50, 58F12, 58F35, 54H20; Secondary 58F25, 54B25.

Library of Congress Cataloging-in-Publication Data

Continuum theory and dynamical systems: proceedings of the AMS-IMS-SIAM joint summer research conference held June 17–23, 1989, with support from the National Science Foundation and the Army Research Office/Morton Brown, editor.
    p. cm.—(Contemporary mathematics, ISSN 0271-4132; 117)
    "The AMS-IMS-SIAM Joint Summer Research Conference in the Mathematical Sciences on Relationships between Continuum Theory and the Theory of Dynamical Systems was held at Humboldt State University, Arcata, California, on June 17–23, 1989"-—T.p. verso.
    Sponsored by the National Science Foundation.
    ISBN 0-8218-5123-3
    1. Differentiable dynamical systems—Congresses. 2. Continuum (Mathematics)—Congresses. I. Brown, Morton, 1931– . II. AMS-IMS-SIAM Joint Summer Research Conference in the Mathematical Sciences on Relationships between Continuum Theory and the Theory of Dynamical Systems (1989: Humboldt State University) III. American Mathematical Society. IV. National Science Foundation (U.S.) V. Series: Contemporary mathematics (American Mathematical Society); v. 117.
QA614.8.C65 1991
515′.352—dc20
                                  91-11451
                                  CIP

Contents

Preface

In the last few years there has been increasing interaction between mathematicians working in dynamical systems and in continuum theory. As a result, a Joint Summer Research Conference on "Relationships between Continuum Theory and the Theory of Dynamical Systems," sponsored by the National Science Foundation was held at Humboldt State University during the week June 18–23, 1989. It was organized by Marcy Barge and Tom Ingram, along with Mort Brown, Bob Devaney, and Bob Williams.

The papers in this volume are representative of the topics covered at the conference. Most are concerned with the dynamics of surface homeomorphisms or of continua that occur as attractors for surface homeomorphisms. They range from "topological" dynamical systems with little reference to continua theory, to the reverse, but on the whole they illustrate the increasing confluence of the two disciplines.

Some papers in this volume are in final form, others are preliminary. Some are expository. There is a concluding section of research problems.

Morton Brown

Contemporary Mathematics
Volume 117, 1991

Whitney's Regular Families of Curves Revisited

JAN M. AARTS AND LEX G. OVERSTEEGEN

ABSTRACT. A separable and metrizable space X is called a flowbox manifold if there exists a base for the open sets each of whose elements has a product structure with the reals, $\mathbf{R}$, as a factor such that a natural consistency condition is met. We show how flowbox manifolds can be divided into orientable and nonorientable ones. We prove that a space X is an orientable flowbox manifold if and only if X can be endowed with the structure of a flow without restpoints. In this way we generalize Whitney's theory of regular families of curves so as to include self-entwined curves in general separable metric spaces.

All spaces under consideration are separable and metrizable.

1. Introduction

Whitney's paper "Regular families of curves"[W] has been of great influence. It laid the groundwork for the theory of cross-sections in flows and it can also be seen as a prelude to the theory of foliations. In Whitney's terminology, a curve is a topological copy of an interval (open, closed, or half-open) or a circle. A family of curves is a partition of a separable metric space into curves. The family is called regular if for every $\epsilon > 0$ and every point p there is a $\delta > 0$ such that, whenever $d(p, q) < \delta$, arbitrarily long arcs that are contained in the curve through p can be pushed onto a similar arc through q by a homeomorphism that moves the points of the arc not more than ϵ. Using these two properties, Whitney proved the existence of cross-sections. He also showed that if the family of curves is orientable and if the space is locally compact, it is possible to endow the space with the structure of a flow in such a way that the open curves and the endpoints of curves coincide with the orbits of the flow.

In this paper we announce a generalization in two directions of these

---

1980 *Mathematics Subject Classification* (1985 *Revision*). Primary 54H20, 54E99.
Key words and phrases. Flow, regular family of curves, flowbox.
The detailed version of this paper, entitled "Flowbox manifolds," has been submitted for publication elsewhere.
The second author was supported in part by NSF-RII-8610669 and NSF-DMS-8602400.

results. First we allow the curves to be self-entwined and only require that there is a continuous bijection of $\mathbf{R}$ (or S^1) onto the curve. Secondly, we perform the construction of a flow for any separable metric space. It seems that our proof has the additional benefit of being less complicated.

We found it more convenient to start our theory with the discussion of the local product structure instead of beginning with regular families of curves. This is formalized in the definition of a flowbox manifold, a notion that encompasses the notions of dimension-one foliation and flow. A flowbox manifold is a space in which every point has arbitrarily small neighborhoods with a product structure in such a way that a natural consistency condition is satisfied (see §2).

MAIN THEOREM. *Let X be an orientable flowbox manifold. Then there is a flow on X such that each streamline in X is contained in some orbit of the flow.*

The converse of the main theorem is well known.

2. Flowbox manifolds

We discuss the basic properties of flowbox manifolds. We will use $\mathbf{R}$ to denote the space of real numbers.

DEFINITION. A separable metric space X is called a *flowbox manifold* if it has the following two properties:

(1) *Local product structure.* There exists a base $\mathcal{U} = \{U_\beta | \beta \in B\}$ for the open subsets such that for each $\beta \in B$ there exists a space S_β and a homeomorphism $h_\beta : S_\beta \times \mathbf{R} \to U_\beta$.

(2) *Consistency.* Suppose that $U_\alpha = h_\alpha(S_\alpha \times \mathbf{R})$ and $U_\beta = h_\beta(S_\beta \times \mathbf{R})$ are elements of $\mathcal{U}$. If $U_\alpha \subset U_\beta$, then for each $s \in S_\alpha$ there exists $t \in S_\beta$ such that $h_\alpha(\{s\} \times \mathbf{R}) \subset h_\beta(\{t\} \times \mathbf{R})$.

Easy examples of flowbox manifolds include foliations [CN], flows [NS], and matchbox manifolds [AM].

STANDING NOTATION. Let S be any space. We define $F_S = S \times [-1, 1]$ and $E_S = S \times (-1, 1)$. The space F_S is called a *standard flowbox*. For each $x \in S$, the set $\{x\} \times [-1, 1]$ is called a *streamline* of F_S. The natural projections of F_S onto S and $[-1, 1]$ are denoted by pr_1 and pr_2, respectively. Both pr_1 and pr_2 are open. As $[-1, 1]$ is compact, pr_1 is closed as well.

We now define flowboxes, which play a very important role in the development of the theory.

DEFINITIONS. Let U_β be an open set witnessing the local product structure of the flowbox manifold X, i.e., $U_\beta = h_\beta(S_\beta \times \mathbf{R})$ for some $\beta \in B$. A closed subset V of X, which is contained in U_β, is called a *flowbox* if there exists a space S, a dense subspace S^0 of S, and a topological embedding

$\phi : F_S \to X$ such that the following holds:

(1) $V = \phi(F_S)$ and $\text{int}_X V = \phi(E_S o)$;

(2)   for each $y \in S$ there is a $z \in S_\beta$ with $\phi(\{y\} \times [-1, 1]) \subset h_\beta(\{z\} \times \mathbf{R})$.

For each point $p \in \text{int}_X V$, the set V is also called a *flowbox neighborhood of p*. The induced map $\phi : F_S \to V$ is called a *parameterization of V*. The sets $\phi(\{y\} \times [-1, 1])$, $y \in S$, are called *streamlines of V* or of X.

It is to be noted that every flowbox V of a flowbox manifold X is a regular closed set, i.e. $V = \text{cl}_X \text{int}_X V$. The following proposition shows that the flowbox neighborhoods form a base for the topology.

PROPOSITION AND DEFINITION. Let X be a flowbox manifold. Then there exists a countable collection $\mathcal{V} = \{V_i | i = 1, 2, \ldots\}$ of flowboxes $V_i = \phi_i(F_{S_i})$ such that the collection $\{\phi_i(E_{S_i^o}) | i = 1, 2, \ldots\}$ is a base for the open sets. Such a base $\mathcal{V}$ is called a *flowbox base of* X.

3. Orientation

In the process of defining a flow structure on a flowbox manifold, we have to single out the sets which will become the orbits.

DEFINITION. Let J be an arc in a flowbox manifold X and let $g : [0, 1] \to J$ be a topological embedding. Then J is called a *partial orbit* of X if for every $t \in (0, 1)$ there exists a flowbox neighborhood $V = \phi(F_S)$ of $g(t)$ such that $V \cap J = \phi(\{x\} \times [-1, 1])$ for some (unique) $x \in S$.

The following proposition is evident.

PROPOSITION. *Every streamline is a partial orbit.*

By analyzing the intersection of a partial orbit and a flowbox one can show:

PROPOSITION. *Suppose that J_1 and J_2 are partial orbits in a flowbox manifold X. If $J_1 \cap J_2 \neq \varnothing$, then $J_1 \cup J_2$ is a partial orbit.*

By the proposition, the following definition is justified.

DEFINITION. Let x be a point in a flowbox manifold. The *orbit of x*, denoted by Γ_x, is the union of all partial orbits containing x.

It follows that the orbits Γ_x form a partition of X into orbits. From the results in [A1] and [AM] it follows that each orbit in a flowbox manifold is either a topological circle, a topological copy of the reals $\mathbf{R}$, or a special one-to-one continuous image of the reals, called a *P-manifold*. We recall the following definition from [A1].

DEFINITION. Let Γ be an orbit in a flowbox manifold. If Γ is a topological circle, then any covering map $p : \mathbf{R} \to \Gamma$ is called a *parameterization of Γ*. If Γ is not a circle, then any continuous bijection $p : \mathbf{R} \to \Gamma$ is called a *parameterization of Γ*.

It turns out that if p_1 and p_2 are parameterizations of an orbit Γ, then the map $p_1^{-1} \circ p_2 : \mathbf{R} \to \mathbf{R}$ is a homeomorphism. If this homeomorphism is increasing, we say that p_1 and p_2 have the same *direction*. Thus there are two directions for each orbit. See [A1, A2] for more details.

DEFINITIONS. Let X be a flowbox manifold. Let $\{\Gamma_\alpha | \alpha \in A\}$ be the collection of orbits of X. If for each $\alpha \in A$ a parameterization $p_\alpha : \mathbf{R} \to \Gamma_\alpha$ is given, we call the collection $\{p_\alpha | \alpha \in A\}$ a *parameterization of* X. If a parameterization has been given, we shall say that a flowbox $V = \phi(F_S)$ is *coherently directed* if for each $x \in S$ and for any closed interval J in $\mathbf{R}$ such that

$$p_\alpha(J) = \phi(\{x\} \times [-1, 1])$$

for some α, the composition $\mathrm{pr}_2 \circ \phi^{-1} \circ p_\alpha$ is increasing.

DEFINITIONS. A flowbox manifold X is said to be *orientable* if there is a parameterization $\{p_\alpha | \alpha \in A\}$ of X such that each point of X has a flowbox neighborhood which is coherently directed. In that case the parameterization is called *proper*. Notice that in an orientable flowbox manifold the parameterization $\{p_\alpha | \alpha \in A\}$ induces an order (also denoted by $<$) on the orbits Γ_α.

4. Main theorem

In this section we will outline how to define a flow on an orientable flowbox manifold.

NOTATION. Throughout this section X is a fixed orientable flowbox manifold. $\{\Gamma_\alpha \mid \alpha \in A\}$ denotes the collection of orbits. For each $\alpha \in A$, $p_\alpha : \mathbf{R} \to \Gamma_\alpha$ is a parameterization of Γ_α and the parameterization $\{p_\alpha \mid \alpha \in A\}$ is proper.

$\mathcal{V} = \{V_i \mid i = 1, 2, \ldots\}$ is a fixed flowbox base for X. We assume that each $V_i = \phi_i(F_{S_i})$ is coherently directed by ϕ_i, $i = 1, 2, \ldots$. For $i = 1, 2, \ldots$ we denote by $\mathcal{C}_i$ the collection of all partial orbits of the form

$$\phi_i(\{x\} \times [a, b]), \qquad x \in S_i^0, \; [a, b] \subset (-1, 1).$$

We write $\mathcal{C} = \bigcup \{\mathcal{C}_i \mid i = 1, 2, \ldots\}$. $\mathcal{C}$ is endowed with the Hausdorff metric. It is to be noted that each $\mathcal{C}_i$ is an open subset of $\mathcal{C}$, $i = 1, 2, \ldots$. If $J \in \mathcal{C}$, then a partition of J is a collection $\{J_1, \ldots, J_k\}$ of elements of $\mathcal{C}$ such that $J = J_1 \cup \cdots \cup J_k$ and $| J_i \cap J_{i+1} | = 1$, $i = 1, \ldots, k - 1$.

LEMMA AND DEFINITION. *For each* $i = 1, 2, \ldots$ *there is a function* $\mu_i : \mathcal{C} \to \mathbf{R}$ *with the following properties.*

(1) μ_i *is supported by* V_i:
   $\mu_i(J) > 0$ *for each* $J \in \mathcal{C}$ *such that* $J \cap \mathrm{int}_X V_i \neq \varnothing$;
   $\mu_i(J) = 0$ *for each* $J \in \mathcal{C}$ *such that* $J \cap \mathrm{int}_X V_i = \varnothing$.
(2) *For each* $J \in \mathcal{C}_1 \cup \cdots \cup \mathcal{C}_i$ *we have* $0 \leq \mu_i(J) \leq 1$.
(3) μ_i *is additive on* $\mathcal{C}$:
   if J_1, J_2 *and* $J_1 \cup J_2$ *belong to* $\mathcal{C}$ *and* $|J_1 \cap J_2| = 1$,
   then $\mu_i(J_1 \cup J_2) = \mu_i(J_1) + \mu_i(J_2)$.
(4) μ_i *is continuous on* $\mathcal{C}$.

In order to define the functions μ_i, first define functions g_i positive on V_i and zero on $X \setminus V_i$. Integrating g_i over the streamlines of V_i gives a function ν_i satisfying (1), (3), and (4) on subarcs of streamlines of V. By partitioning an arbitrary arc $J \in \mathcal{C}$ into subarcs J_k such that each arc J_k is either a streamline of V_i, or $J_k \cap \text{Int}_X V_i = \varnothing$, one can extend the definition of ν_i over all of $\mathcal{C}$. In order to satisfy (2), in addition, one needs to analyze the intersection of V_i with V_j ($j < i$) and accordingly choose the functions g_i sufficiently small.

PROOF OF THE MAIN THEOREM. We use the notation of the lemma.

The function $\mu : \mathcal{C} \to \mathbf{R}$ is defined by

$$\mu(J) = \sum_{k=1}^{\infty} \frac{\mu_k(J)}{2^k}.$$

It is to be noted that if $J \in \mathcal{C}_i$, then $\mu_k(J) \leq 1$ for all $k \geq i$ by condition 2 of the lemma, whence $\sum_{k=i}^{\infty} \mu_k(J)/2^k \leq 2^{-i+1}$. It follows that the series at the right-hand side is uniformly convergent on each $\mathcal{C}_i$, $i = 1, 2, \ldots$. It is now clear that μ is continuous and additive on $\mathcal{C}$.

One can think of μ as a time span function, that is, $\mu(J)$ measures the time it takes to travel from one endpoint of J to the other. In order to define a flow on X, we will first define a germ of a local flow as follows:

Let $x \in X$ and let V be a flowbox neighborhood of x. Since X is an orientable flowbox manifold, each orbit inherits a natural order $<$ from $\mathbf{R}$. For $t > 0$ (sufficiently small), there exists a unique point y in the orbit of x such that $\mu([x, y]) = t$ and $x < y$. Define $\pi(x, t) = y$. In this way one can define a germ of a local flow. There is a standard way of extending the domain of π in order to obtain a local flow, also denoted by π (see [H] for details). The proof is completed by invoking a result of Carlson [C] which states that for a local flow there exists a reparameterization which turns it into an equivalent global flow. It follows from a result of Ura [U] that this flow is unique up to a reparameterization.

5. Regular families of curves

In Whitney's paper [W] a family of curves in a space X was defined as a partition of X into arcs, with or without endpoints, and circles. In the second part of the paper, arcs with endpoints are excluded. We shall extend Whitney's major results in such a way that more general curves are admitted.

DEFINITION. A *curve* is either a circle or a separable metric space which is a one-to-one continuous image of the real line $\mathbf{R}$. A *family of curves* in a separable metrizable space X is a partition $\{C_\alpha \mid \alpha \in A\}$ of the space into curves.

REMARK. In what follows it is convenient to denote a family of curves in X by $\{C_x \mid x \in X\}$ where $x \in C_x$ for each $x \in X$. So every element of the partition is labeled by each of its elements.

DEFINITION. A family $\{C_x \mid x \in X\}$ of curves in X is called *regular* if for each $\epsilon > 0$ and for each $x \in X$ and for each arc J with $x \in J \subset C_x$, there exists a $\delta > 0$ such that for every $y \in B_\delta(x)$ there is an ϵ-embedding $h : J \to C_y$ (i.e., $d(s, h(s)) < \epsilon$ for every $s \in J$) such that $h(x) = y$.

The following argument is an outline of the equivalence of the notions of flowbox manifold and regular family of curves. It is similar to the argument designed by Whitney in the original situation [W].

DEFINITION. Suppose that $\{C_x \mid x \in X\}$ is a regular family of curves in X. Let $x \in X$. A closed set S is a *local section at* x provided that $x \in S$ and that there exists a neighborhood U of x such that for each $y \in \mathrm{cl}_X U$ each component of $C_y \cap \mathrm{cl}_X U$ intersects S in exactly one point.

LEMMA. *Suppose that* $\{C_x \mid x \in X\}$ *is a regular family of curves in* X. *Then* X *admits a local section at every point* $x \in X$.

THEOREM. *Suppose that* $\{C_x \mid x \in X\}$ *is a regular family of curves in* X. *Then for each* $x \in X$ *there exists a neighborhood* U *of* x, *a section* S *at* x, *and a topological embedding* $\psi : S \times [-1, 1] \to \mathrm{cl}_X U$ *such that*

(i) $\psi(S \times \{0\}) = S$ *and* $\psi(x, 0) = x$,
(ii) $\psi(s, t) \in C_s$ *for each* $s \in S$ *and* $t \in [-1, 1]$.

THEOREM. *A separable metrizable space* X *is a flowbox manifold if and only if there is a regular family of curves in* X.

PROOF. The "if" part immediately follows from the preceding theorem. It is to be noted that the consistency condition follows from the fact that we start out with a family of curves by which the local products are lined up.

To prove the "only if" part we show that if X is a flowbox manifold, then the family $\{\Gamma_x \mid x \in X\}$ of all orbits is a regular family. If the flowbox manifold X is orientable, then by the main theorem there is a flow $\pi : X \times \mathbf{R} \to X$ such that each streamline of X is contained in some orbit of the flow. It easily follows that every C_x coincides with some orbit in the flow. The condition expressed in the definition of a regular family is a well-known property in the theory of flows. Sometimes it is called the continuity of the initial conditions [NS]. If the flowbox manifold is not orientable we have to fall back on the method of pasting together of flowboxes. This method has been described in a very detailed way in [A1] for a P-manifold and can be adapted for the present situation with only minor modifications.

REFERENCES

[A1] J. M. Aarts, *The structure of orbits in dynamical systems*, Fund. Math. **129** (1988), 39–58.
[A2] ———, *Orientations of orbits in flows*, Ann. New York Acad. Sci. **552**, 1–7.
[AM] J. M. Aarts and M. Martens, *Flows on one-dimensional spaces*, Fund. Math. **131** (1988), 53–67.
[CN] C. Camacho and A. L. Neto, *Geometric theory of foliations*, Birkhauser, Basel, 1985.
[Ca] D. H. Carlson, *A generalization of Vinograd's theorem for dynamical systems*, J. Differential Equations **11** (1972), 193–201.
[H]   O. Hajek, *Local characterization of local semi-dynamical systems*, Math. Systems Theory **2** (1968), 17–25.

[NS] V. V. Nemytskii and V. V. Stepanov, *Qualitative theory of differential equations*, Princeton Mathematical Series, no. 22, Princeton Univ. Press, Princeton, New Jersey, 1960.

[U]   T. Ura, *Isomorphisms and local characterization of local dynamical systems*, Funkcial. Ekvac. **12** (1969), 99–122.

[W]   H. Whitney, *Regular families of curves*, Ann. of Math. **34** (1933), 244–270.

TECHNISCHE UNIVERSITEIT DELFT, FACULTEIT DER WISKUNDE EN INFORMATIKA, POSTBUS 356, 2600 A J DELFT, THE NETHERLANDS

UNIVERSITY OF ALABAMA AT BIRMINGHAM, DEPARTMENT OF MATHEMATICS, BIRMINGHAM, ALABAMA 35294, U.S.A.

Contemporary Mathematics
Volume 117, 1991

Sets of Periodic Points of Functions on Trees

STEWART BALDWIN

Let $\leq_s$ (the Sarkovskii ordering) be the following linear ordering of the positive integers:

$$1 \leq_s 2 \leq_s 2^2 \leq_s 2^3 \leq_s \cdots 5 \cdot 2^2 \leq_s 3 \cdot 2^2$$
$$\leq_s \cdots 9 \cdot 2 \leq_s 7 \cdot 2 \leq_s 5 \cdot 2 \leq_s 3 \cdot 2 \leq_s \cdots 9 \leq_s 7 \leq_s 5 \leq_s 3.$$

(A more precise definition will be given below.)

SARKOVSKII'S THEOREM. *Let I be the unit interval $[0, 1]$. Then for all positive integers k and m, $m \leq_s k$ if and only if for every continuous $f: I \to I$, if f has a point of period k, then f also has a point of period m. (The theorem is also true if I is replaced by the set of all real numbers.)*

DEFINITION. Let I be the unit interval $[0, 1]$ and R the set of real numbers. The *n-od* (often called X_n below) is the set of all complex numbers z such that z^n is in the unit interval I, i.e., a central point with n copies of I attached. The 3-od is called the *triod*. A *graph* is any subset of R^n which is the union of finitely many compact straight line segments, and a *tree* is any connected graph which contains no homeomorphic copy of the circle (or, equivalently, a graph which is uniquely arcwise connected). If $f: X \to X$ is a function, then a point x of X will be called *periodic with respect to f* if $f^n(x) = x$ for some integer $n > 0$, where f^n is f composed with itself n times. We call the least such n the *period* of x. We define Per(f) to be the set $\{n : f \text{ has a point of period } n\}$.

Attempts to find generalizations of Sarkovskii's Theorem have turned in various directions. The generalization discussed here follows a line which was first suggested by Alseda, Llibre, and Misiurewicz in [ALM], where it was pointed out that Sarkovskii's Theorem can be viewed as a characterization of all sets Per(f), where f ranges over continuous functions on $[0, 1]$. Thus, given a topological space X, we would like to describe the set Per(f),

---

1980 *Mathematics Subject Classification* (1985 *Revision*). Primary 58F20.
The final version of this paper will be submitted for publication elsewhere.

where f is a continuous function on X. This was solved in [**ALM**] for all functions of the triod leaving the central point fixed, and in [**B**] for all continuous functions on the n-od. Since the results given in [**B**], the problem has been successfully solved for all trees having two nodes, and we describe these results in this preliminary report. The full details will appear elsewhere. We first describe the main result of [**B**].

DEFINITION. For each positive integer n, define partial orderings $\leq_n$, where the domain of the relation $\leq_n$ is the set $\{k : k$ an integer, $k = 1$ or $k \geq n\}$. The ordering $\leq_1$ is defined by $2^i \leq_1 2^{i+1} \leq_1 2^{j+1}(2m+1)$ $\leq_1 2^j(2k+3) \leq_1 2^j(2k+1)$, for all integers $i, j \geq 0$ and $k, m > 0$ (i.e., $\leq_1$ is just the Sarkovskii ordering $\leq_s$). If $n > 1$ then the ordering $\leq_n$ is defined as follows. Let m, k be positive integers.

CASE 1. $k = 1$. Then $m \leq_n k$ iff $m = 1$.

CASE 2. k is divisible by n. Then $m \leq_n k$ iff either $m = 1$ or m is divisible by n and $(m/n) \leq_1 (k/n)$.

CASE 3. $k > 1$, k not divisible by n. Then $m \leq_n k$ iff either $m = 1$, $m = k$, or $m = ik + jn$ for some integers $i \geq 0$, $j \geq 1$.

Note that, unlike our treatment in [**B**], the integers $\{2, 3, \ldots, n-1\}$ are not included in the domain of the $\leq_n$ ordering. In [**B**], it did not make any difference whether or not these integers were included in the ordering. For the generalizations given here, these integers need to be omitted.

A set Z of positive integers is an *initial segment* of $\leq_n$ if whenever k is a member of Z and $m \leq_n k$ then m is also a member of Z, i.e., Z is closed under $\leq_n$-predecessors. The following theorem was the main theorem of [**B**].

THEOREM. *Let f be a continuous function on the n-od. Then $\mathrm{Per}(f)$ is a nonempty finite union of initial segments of $\{\leq_p : 1 \leq p \leq n\}$. Conversely, if Z is a nonempty finite union of initial segments of $\{\leq_p : 1 \leq p \leq n\}$, then there is a continuous function f on the n-od which fixes the central point 0 such that $\mathrm{Per}(f) = Z$. (In taking the above union, we use the usual notation that identifies a relation with a set of ordered pairs in the obvious way.)*

To generalize this theorem to larger trees, more orderings are needed. First, we give a few examples of maps which illustrate the new basic ideas for trees having two nodes. Let H be the tree which is obtained by joining two triods at their endpoints (and is shaped like the capital letter H), and we refer the reader to Figure 1. In each case in the figure, a and b (the nodes) are points of period two which map to each other, and the other points form an orbit in the indicated way, with 1 being mapped to 2, which maps to 3, and so forth up to the largest integer shown, which maps back to 1. The remaining points are mapped in a piecewise linear way. The first function has points of periods 1, 2, and 6, and no other periodic points. The other two functions are the first two in an infinite family of functions f such that $\mathrm{Per}(f) = A - \{4\}$,

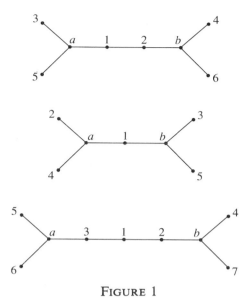

FIGURE 1

where A is some initial segment of the Sarkovskii ordering, and illustrate why 2 should be left out of the $\leq_3$ ordering when the $\leq_{(3,2)}$ ordering is described below. (The relevant theorem in [B] is Theorem 2.11, which needs to be modified in this more general setting, as the examples given here show.) We leave it to the reader to go through the easy steps of drawing the Markov graph of the functions to see that the above assertions are correct.

Since the problem (at the time this is being written) has only been solved for trees having two or fewer nodes, the following definition may not be relevant to further generalizations. However, it is the definition needed for the cases already solved, and the arguments used suggest that these orderings will be relevant for periodic points on larger trees.

DEFINITION. For all ordered pairs (m, n), define the ordering $\leq_{(m,n)}$. The domain of the relation $\leq_{(m,n)}$ will be the set $\{k : k$ an integer, $k = 1$ or $k > n(m-1)\}$.

CASE 1. $k = 1$. Then $j \leq_{(m,n)} k$ iff $j = 1$.

CASE 2. k is divisible by n. Then $j \leq_{(m,n)} k$ iff either $j = 1$ or j is divisible by n and $(j/n) \leq_m (k/n)$.

CASE 3. $k > 1$, k not divisible by n. Then $j \leq_{(m,n)} k$ iff either $j = 1$, $j = k$, or $j = ik + pn$ for some integers $i \geq 0$, $p > 1$.

To visualize what this ordering looks like in the case $n = 2$, it is easiest to divide it into three pieces. At the top is all of the odd numbers greater than $2m - 2$, arranged in descending order, as in the Sarkovskii ordering. At the bottom of the ordering is 1. In between is the set of all even integers greater than $2m - 2$, with the ordering induced by the $\leq_m$ ordering by multiplying all of the integers in the $\leq_m$ ordering by 2.

DEFINITION. For each pair (m, n) the tree $X_{(m,n)}$ is defined to be n copies of the m-od attached at endpoints (so $X_{(3,2)}$ is the tree H described above). If S and T are trees, then we define $S \le T$ iff a homeomorphic copy of S can be obtained from T by shrinking a finite number of closed connected subsets of T to points.

MAIN THEOREM. *Let T be a tree with two nodes, and let f be a continuous function on T. Then $\operatorname{Per}(f)$ is a union of initial segments of the orderings $\{\le_{(m,n)} : X_{(m,n)} \le T\} \cup \{\le_m : X_m \le T\}$. Conversely, if A is a nonempty set of integers which is a union of initial segments of $\{\le_{(m,n)} : X_{(m,n)} \le T\} \cup \{\le_m : X_m \le T\}$, then there is a continuous function f on T taking nodes to nodes such that $\operatorname{Per}(f) = A$.*

Partial results are known for many larger trees, but a complete result such as the above theorem is not yet available. At the present state of knowledge there seem to be two main difficulties, and it is not clear whether it is just a matter of working out the details using the methods already developed, or whether new techniques are required. Roughly speaking, the two difficulties are as follows:

1. When the "conversely" part of the main theorem is proven, the main strategy (as in [B]) is to produce several separate examples and then "glue" them together. This was easy in the case of the n-od, and can be done for two node trees with a little more effort, but the problem of attaching two examples together on larger trees (without introducing any unwanted new periodic points) presents difficulties.

2. The smallest two trees that are incomparable with respect to the ordering $\le$ on trees given above are the 5-od and the tree H mentioned above. Let T be the tree that is obtained by attaching H and the 5-od at two endpoints. Then T (and anything larger than it in the $\le$ ordering on trees) presents an additional problem, for suppose f is a continuous function on T sending H to the 5-od and the 5-od to H. Then f^2 maps H into itself, and we are essentially confronted with a problem of the following type: Given X and a continuous map f on X, what can we say about $\operatorname{Per}(f)$ if we know that it factors through some other tree, i.e., there is a tree Y and functions $g: X \to Y$ and $h: Y \to X$ such that $hg = f$. For example, if Y is an arc, then it is easy to see that $\operatorname{Per}(f)$ must be an initial segment of the Sarkovskii ordering (regardless of what X is), but if X and Y are trees which cannot be compared by the $\le$ ordering, the case is not so clear.

BIBLIOGRAPHY

[ALM] L. Alseda, J. Llibre, and M. Misiurewicz, *Periodic orbits of maps of Y*, preprint, 1987.
  [B] S. Baldwin, *An extension of Sarkovskii's Theorem to the n-od*, preprint, 1988 (submitted to Ergodic Theory and Dynamical Systems).

AUBURN UNIVERSITY, AUBURN, ALABAMA 36849-5310

Contemporary Mathematics
Volume **117**, 1991

Indecomposability and Dynamics
of Invariant Plane Separating Continua

MARCY BARGE AND RICHARD M. GILLETTE

1. Introduction

In his fundamental study of twist maps of the annulus, Birkhoff [**Bi**] observed that complicated dynamical properties of an annulus homeomorphism may necessarily lead to a complicated topological structure for certain invariant subsets. This is the case for his "remarkable curve"—an invariant plane separating continuum. In 1934 Charpentier [**Ch**] proved that the Birkhoff continuum is in fact an indecomposable continuum (a continuum that is not the union of two proper subcontinua).

Another instance of complicated dynamics inducing complicated topology occurs in Cartwright and Littlewood's investigation of solutions to the forced van der Pol equations [**CL1, CL2**]. They found that (at certain parameter values in the van der Pol equation) an associated Poincaré homeomorphism admits an invariant plane separating continuum containing periodic points of consecutive odd periods and containing no fixed points. Cartwright and Littlewood conjectured that this continuum must contain an indecomposable subcontinuum (see the remarks preceding Theorem 7 of [**CL2**]). We prove two theorems that apply to the above examples and that give substance to the correspondence between complicated dynamical properties of plane homeomorphisms and exotic topology in invariant plane separating continua.

We consider an orientation-preserving homeomorphism of the plane that leaves invariant a continuum Λ. We will assume that Λ is an irreducible plane-separating continuum (Λ separates the plane and no proper subcontinuum of Λ has this property) and that the complement of Λ consists of exactly two domains. Such a continuum will be called a cofrontier as it is necessarily the frontier of each of its complementary domains. Our theorems assert that, under certain dynamical hypotheses on the homeomorphism, an

1980 *Mathematics Subject Classification* (1985 *Revision*). Primary 54H20.

This paper is in final form and no version of it will be submitted for publication elsewhere.

invariant cofrontier must be an indecomposable continuum. Such continua have a complicated structure and possess many remarkable properties (see, for example, [K1]).

If a plane homeomorphism leaves invariant a plane-separating continuum M, it is not necessarily the case that M contains an invariant irreducible plane separating continuum (see the example in [Kn]). If, however, M has finitely many complementary domains then M does contain an irreducible plane-separating continuum [K2] and it is not difficult to argue that if, in addition, M has empty interior then M must contain an irreducible plane-separating continuum invariant under some power of the homeomorphism. Kuratowski also demonstrated [K2] that an irreducible plane-separating continuum that separates the plane into three or more domains (examples of this are the lakes of Wada continuum [HY] and the Plykin attractor [P]) is necessarily either indecomposable or the union of two indecomposable subcontinua. It is for this reason that we restrict our attention to cofrontiers.

By means of the Caratheodory theory of prime ends [C], one may associate two rotation numbers, ρ_i and ρ_e, to the orientation-preserving plane homeomorphism F and its invariant cofrontier Λ. (These are the rotation numbers of the homeomorphisms induced by F on the circles of prime ends associated with the bounded and unbounded complementary domains of Λ.) Theorem A states that if one of these rotation numbers is nonzero and if F has a fixed point in Λ, then Λ is an indecomposable continuum. We obtain as a corollary that if one of ρ_i and ρ_e is rational and the other is irrational or if both ρ_i and ρ_e are rational with different denominators (in reduced form) then Λ is an indecomposable continuum. This corollary is a special case of the result of Charpentier [Ch] that Λ is indecomposable provided $\rho_i \neq \rho_e$. For a very nice presentation of the Charpentier result, and a development of other properties of what are now called Birkhoff attractors, see Le Calvez [LeC].

Theorem B states that if the orientation-preserving homeomorphism F has periodic points of periods m and n in the invariant cofrontier Λ, with m and n relatively prime, and if F has no fixed points in Λ, then Λ is an indecomposable continuum. As a corollary to this we prove that the continuum, denoted by F^* in [CL2], arising in the van der Pol equations and conjectured by Cartwright and Littlewood to contain an indecomposable continuum is, in fact, indecomposable.

It has been pointed out to us that the results obtained here can also be gotten using the methods of Le Calvez [LeC]. For yet a different approach to these and other results, see [BG].

In the next section we give the necessary definitions, state our results, and give some examples. In §§III(A), (B), and (C) we will develop (or refer the reader to) needed results on the topology of cofrontiers, the theory of prime ends, and Alexander-Spanier cohomology, respectively. The theorems and their corollaries will be proved in §III(D).

II. Statement of results

By a *continuum* we will mean a nonempty compact and connected metric space. In order to streamline the language, we have included in this definition the (usually excluded) case of a single point. A continuum is *indecomposable* if it is not the union of two proper subcontinua. If Λ is a continuum in the plane we will say that Λ is a *cofrontier* if Λ separates the plane into exactly two domains and no proper subcontinuum of Λ separates the plane. We will denote the complementary domains of a cofrontier by $\mathscr{D}_e$ (the exterior, or unbounded, domain) and $\mathscr{D}_i$ (the interior, or bounded, domain).

Associated with each of the complementary domains $\mathscr{D}_e$ and $\mathscr{D}_i$ of a cofrontier Λ there are *prime points* and *prime ends*, defined originally by Caratheodory [C]. Our terminology (and most of our notation) is borrowed from Mather's treatment [M]. A *chain* in $\mathscr{D}_i$ is a sequence $\{V_n\}$ of open connected subsets of $\mathscr{D}_i$ such that $V_1 \supset V_2 \supset \cdots$; $\mathscr{F}_{\mathscr{D}_i} V_n$ (the frontier of V_n in $\mathscr{D}_i$) is nonempty and connected for each n; and $\mathrm{cl}(\mathscr{F}_{\mathscr{D}_i} V_n) \cap \mathrm{cl}(\mathscr{F}_{\mathscr{D}_i} V_m) = \varnothing$ for $n \neq m$ (here cl denotes the closure in the plane $\mathbb{R}^2$). The chain $\tau = \{W_n\}$ is said to *divide* the chain $\sigma = \{V_n\}$ provided for each n there is an m such that $W_m \subset V_n$. Two chains are *equivalent* if each divides the other, and a chain is *prime* if every chain which divides it is equivalent to it. A *prime point* of $\mathscr{D}_i$ is an equivalence class of prime chains of $\mathscr{D}_i$ and a *prime end* of $\mathscr{D}_i$ is a prime point of $\mathscr{D}_i$ with a representation prime chain $\{V_n\}$ such that $\mathscr{F} V_n \cap \Lambda \neq \varnothing$ for all n.

We will let $\hat{\mathscr{D}}_i$ denote the collection of all prime points of $\mathscr{D}_i$. Given an open set W in $\mathscr{D}_i$ and an element $\mathscr{P}$ of $\hat{\mathscr{D}}_i$ we will say that $\mathscr{P}$ *divides* W provided a representative chain $\{V_n\}$ of $\mathscr{P}$ has the property that $V_n \subset W$ for some n. We let $\tilde{W} = \{\mathscr{P} \in \hat{\mathscr{D}}_i | \mathscr{P} \text{ divides } W\}$. The collection of all $\tilde{W}$ for W open in $\mathscr{D}_i$ then defines a basis for a topology on $\hat{\mathscr{D}}_i$. Convergence in this topology works as follows: $\lim_{n \to \infty} \mathscr{P}_n = \mathscr{P}$ if and only if, given a representative chain $\{V_n\}$ of $\mathscr{P}$, for each m there is an N such that $\mathscr{P}_n$ divides V_m for all $n \geq N$. With this topology, $\hat{\mathscr{D}}_i$ is homeomorphic with a closed disk, and the collection of prime ends in $\hat{\mathscr{D}}_i$ is homeomorphic with a circle.

If the plane is compactified by adding a point at ∞ then $\mathscr{D}_e \cup \{\infty\}$ becomes an open disk and, defining prime points and prime ends of $\mathscr{D}_e \cup \{\infty\}$ and topologizing them, as above, we obtain a space, which we denote by $\hat{\mathscr{D}}_e$, that is also homeomorphic with a closed disk. Furthermore, the collection of prime ends in $\hat{\mathscr{D}}_e$ is homeomorphic with a circle. We will denote the circles of prime ends in $\hat{\mathscr{D}}_i$ and $\hat{\mathscr{D}}_e$ by $\mathscr{S}_i$ and $\mathscr{S}_e$, respectively.

If $F \colon \mathbb{R}^2 \to \mathbb{R}^2$ is an orientation-preserving homeomorphism and Λ is an invariant cofrontier ($F(\Lambda) = \Lambda$) then $\mathscr{D}_i$ and $\mathscr{D}_e$ are invariant under F. $\mathscr{S}_e$ and $\mathscr{S}_i$ inherit orientations from the plane and F induces orientation-preserving homeomorphisms $\hat{F}_i \colon \mathscr{S}_i \to \mathscr{S}_i$ and $\hat{F}_e \colon \mathscr{S}_e \to \mathscr{S}_e$ on the circles of prime ends by: given $\mathscr{P}$ in $\mathscr{S}_i$ ($\mathscr{S}_e$) with representative chain $\{V_n\}$ then $\hat{F}_i(\mathscr{P})$ ($\hat{F}_e(\mathscr{P})$) is the prime end with representative chain $\{F(V_n)\}$. The homeomorphisms $\hat{F}_i$ and $\hat{F}_e$ then have *rotation numbers* ρ_i and ρ_e defined as follows. Let $\pi_i \colon \mathbb{R} \to \mathscr{S}_i$ be an orientation-preserving covering projection with period 1. If $f_i \colon \mathbb{R} \to \mathbb{R}$ is a lift of $\hat{F}_i$ then $\rho_i = \lim_{n \to \infty} f_i^n(x)/n$ (mod 1). The limit exists and is independent of $x \in \mathbb{R}$ and of the particular lift f_i (see, for example, [D] or [Ni]). The rotation number ρ_e is defined analogously. We will call ρ_i and ρ_e the *prime end rotation numbers* of F associated with Λ. Proofs of the following theorems and corollaries are deferred to §III D.

THEOREM A. *Suppose F is an orientation-preserving homeomorphism of the plane with invariant cofrontier Λ. If F has a fixed point in Λ and one of the prime end rotation numbers of F associated with Λ is nonzero, then Λ is an indecomposable continuum.*

COROLLARY 1. *Suppose that F is an orientation-preserving homeomorphism of the plane with invariant cofrontier Λ and associated prime end rotation numbers ρ_i and ρ_e.*

- (i) *If one of ρ_i and ρ_e is rational and the other is irrational then Λ is an indecomposable continuum.*
- (ii) *If ρ_i and ρ_e are rational with different denominators (in reduced form) then Λ is an indecomposable continuum.*

It is not difficult to find examples of invariant cofrontiers for which the induced homeomorphisms $\hat{F}_i$ and $\hat{F}_e$ on the circles of prime ends are not topologically conjugate yet in which the cofrontier is hereditarily decomposable [W]. Thus it is really the distinct rotational behaviors of $\hat{F}_i$ and $\hat{F}_e$ that force indecomposability. Also, it is crucial that Λ be a cofrontier, and not just the frontier of a single open disk, as an example at the end of this section will show.

A point p is said to be *periodic of period n* (under F) if $F^n(p) = \underbrace{F \circ \cdots \circ F}_{n \text{ times}}(p) = p$ and $F^k(p) \neq p$ for $1 \le k < n$.

THEOREM B. *Suppose that F is an orientation-preserving homeomorphism of the plane with invariant cofrontier Λ. If F has periodic points of relatively prime periods m and n in Λ, and F has no fixed points in Λ, then Λ is an indecomposable continuum.*

Equivalent to Theorem B is

COROLLARY 2. *Suppose that F is an orientation-preserving homeomorphism of the plane with invariant cofrontier Λ. If F has periodic points of periods m and n in Λ but has no periodic points of period k in Λ for any common divisor k of m and n, then Λ is an indecomposable continuum.*

In [CL1] Cartwright and Littlewood demonstrated that, for certain values of the parameters (large k, b in certain intervals $\mathscr{B}_2$) in the forced van der Pol equation

$$\ddot{y} - k(1 - y^2)\dot{y} + y = b\lambda \cos(\lambda t + a),$$

there is an associated orientation-preserving homeomorphism of the $(y, \dot{y})$-plane (the time $w = 2\pi/\lambda$ map of the flow) possessing stable periodic orbits of periods $2n - 1$ and $2n + 1$ (n is of the order of k). These periodic points lie in an invariant cofrontier, denoted by F^* in [CL2], that contains no fixed points. (F^* is an irreducible subcontinuum contained in the K_0 of [CL1].) A similarity between this situation and that of Birkhoff's "remarkable curve" was noted in [CL1], and in [CL2] Cartwright and Littlewood conjectured that F^* contains an indecomposable subcontinuum (see the discussion accompanying Theorem 7 of [CL2]). It follows from Theorem B that

COROLLARY 3. F^* *is an indecomposable continuum.*

The homeomorphism conisdered by Cartwright and Littlewood is actually a diffeomorphism and the points of periods $2n - 1$ and $2n + 1$ are differentiably hyperbolic attractors. It follows that these periodic points persist under C^1 small perturbations. It can also be checked that the (perturbed) periodic points still lie in an invariant (under the new homeomorphism) cofrontier that also contains no fixed points, provided the perturbation is C^1 sufficiently small. Hence indecomposability of an invariant cofrontier persists, in the example, under C^1 small perturbations. Levi [L] in fact shows that the diffeomorphism is structurally stable so that an invariant cofrontier homeomorphic with F^* persists under C^1 small perturbations.

The continuum F^* also contains a multitude of hyperbolic periodic saddles [CL1, L] which possess rotary homoclinic orbits (see [L], [HH], [AS]). Since the closure of the unstable manifold of each such saddle separates the plane and is contained in F^*, F^* is equal to the closure of each of these unstable manifolds. Indecomposability of F^* thus also follows from [B].

EXAMPLE 1. If m and n are two positive integers and k is the greatest common divisor of m and n then there is an orientation-preserving homeomorphism $F: \mathbb{R}^2 \to \mathbb{R}^2$ having an invariant cofrontier Λ such that $F|_\Lambda$ has periodic points of periods m, n, and k and no other periods, and such that Λ is hereditarily decomposable. For example, if $m = 4$, $n = 6$, $k = 2$, let Λ consist of two line segments, two triods, and four curves spiraling onto these as in Figure 1.

Let F be rotation of the entire plane through π followed by a homeomorphism that rotates one of the triods through $2\pi/3$ about its center, pulling in along the spiraling arcs, and that rotates one of the line segments through π about its center, pulling in along the spiraling arcs.

EXAMPLE 2. Given any two positive integers m, n, there is an orientation-preserving homeomorphism $F: \mathbb{R}^2 \to \mathbb{R}^2$ having the properties: F leaves

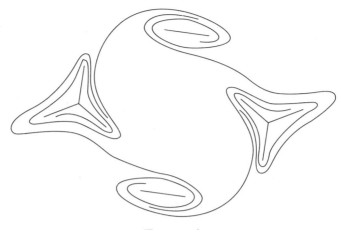

FIGURE 1

invariant the frontier Λ of a connected, simply-connected, bounded domain; $F|_\Lambda$ has periodic points of periods m and n and of no other periods; and Λ is hereditarily decomposable.

For example, if $m = 2$ and $n = 3$, let Λ consist of a circle together with two curves spiraling onto the circle as in Figure 2.

Now let F rotate the circle in Λ through $2\pi/3$ and interchange the two spiraling curves so that their endpoints constitute a periodic orbit of period 2.

Note also that the interior prime end rotation number, ρ_i, of F^3 is $1/2$ and that F^3 has fixed points in Λ. One sees from this the importance of the assumption in Theorem A that Λ is a cofrontier.

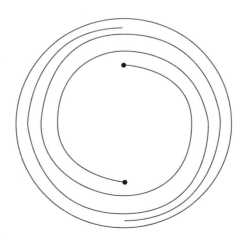

FIGURE 2

III. Preliminaries to the proofs of the theorems

A. Recall that a cofrontier is a continuum Λ that irreducibly separates the plane into exactly two complementary domains. Denoting these domains by $\mathscr{D}_e$ and $\mathscr{D}_i$, we see that, since the frontiers of $\mathscr{D}_e$ and $\mathscr{D}_i$ are plane-separating continua contained in Λ, $\mathscr{F}\mathscr{D}_e = \Lambda = \mathscr{F}\mathscr{D}_i$. Conversely, if Λ is a continuum that separates the plane into exactly two domains, $\mathscr{D}_e$ and $\mathscr{D}_i$, and $\mathscr{F}\mathscr{D}_e = \Lambda = \mathscr{F}\mathscr{D}_i$, then Λ is a cofrontier. A proof of this fact is contained in the proof of Lemma 41 of [**CL2**].

LEMMA A1. *Suppose that H and K are proper subcontinua of the cofrontier Λ. Then*

(i) *$\Lambda - H$ is connected,*

(ii) *if $H \cup K \neq \Lambda$ then $H \cap K$ is connected,*

(iii) *if $H \cup K = \Lambda$ then $H \cap K$ has exactly two components,*

(iv) *if $\Lambda - (H \cup K)$ is not connected then $\Lambda - (H \cup K)$ has exactly two components.*

PROOF. Parts (ii) and (iii) follow from Theorem 11.1 of [**N**] and the fact that H and K do not separate the plane. For (i), assume that $\Lambda - H = A \cup B$ with A and B disjoint and relatively open in $\Lambda - H$. Then $L = A \cup H$ and $M = B \cup H$ are subcontinua of Λ with $\Lambda = L \cup M$ and $L \cap M = H$. Then $L \cap M$ is connected and it follows from (iii) that one of L and M is not proper. Thus either $A = \varnothing$ or $B = \varnothing$ and $\Lambda - H$ is connected.

For (iv), note that if $H \cap K \neq \varnothing$ then $H \cup K$ is a subcontinuum of Λ so that $\Lambda - (H \cup K)$ is connected. Thus $H \cap K = \varnothing$. Now if $\Lambda - (H \cup K)$ has more than two components, there are nonempty, pairwise disjoint, relatively open sets N_1, N_2, and N_3 in $\Lambda - (H \cup K)$ such that $\Lambda - (H \cup K) = N_1 \cup N_2 \cup N_3$. Let $L_i = A \cup B \cup N_i$ for $i = 1, 2, 3$. Then each L_i is closed. If L_1 is connected then $\Lambda - L_1$ is connected (by (i)), but $\Lambda - L_1 = N_2 \cup N_3$. Thus L_1, and similarly L_2 and L_3, are not connected. For each $i = 1, 2, 3$, let C_i be the union of all components of $\Lambda - (H \cup K)$ that are contained in N_i and that have frontier points in H and let D_i be the union of all components of $\Lambda - (H \cup K)$ that are contained in N_i and that have frontier points in K. Let $L_i' = H \cup C_i$ and $L_i'' = K \cup D_i$. Since the frontier of each component of $\Lambda - (H \cup K)$ meets $H \cup K$ [**K1**, Theorem 2 of §47.III] and since each N_i is a union of components of $\Lambda - (H \cup K)$, $L_i = L_i' \cup L_i''$ for $i = 1, 2, 3$. Clearly L_i' and L_i'' are connected and, since L_i is not connected, $L_i' \cap L_i'' = \varnothing$. We now have that

$$\Lambda = \left(\bigcup_{i=1}^{3} L_i' \right) \cup \left(\bigcup_{i=1}^{3} L_i'' \right)$$

is the union of two disjoint subcontinua, in contradiction to the connectedness of Λ. Thus $\Lambda - (H \cup K)$ has exactly two components. $\quad\square$

We will say that the pair $\{A, B\}$ of subcontinua of the cofrontier Λ is an *irreducible separating pair* provided $A \cup B$ separates Λ and that $\{A, B\}$ is irreducible, amongst pairs of subcontinua, with respect to this property.

LEMMA A2. *If H and K are subcontinua of the cofrontier Λ, and $H \cup K$ separates Λ, then there is an irreducible separating pair $\{A, B\}$ with $A \subset H$ and $B \subset K$.*

PROOF. This follows easily from the Brouwer reduction lemma [**HW**, p. 161]. □

Let int_Λ denote the interior relative to Λ.

LEMMA A3. *If $\{A, B\}$ is an irreducible separating pair in the cofrontier Λ then $\text{int}_\Lambda(A) = \varnothing = \text{int}_\Lambda(B)$ and $A \cup B = \text{cl}(U) \cap \text{cl}(V)$, U and V being the components of $\Lambda - (A \cup B)$.*

PROOF. Since $A \cup B$ separates Λ, $\Lambda - (A \cup B)$ is the union of two components U and V (Lemma A1(iv)). If the closure of one of these does not meet one of A or B, say $\text{cl}(U) \cap A = \varnothing$, then U and $A \cup V$ are disjoint relatively open sets (in $\Lambda - B$) whose union is $\Lambda - B$. But $\Lambda - B$ is connected (Lemma A1(i)). Thus $\text{cl}(U) \cap A \neq \varnothing \neq \text{cl}(V) \cap A$ and $\text{cl}(U) \cap B \neq \varnothing \neq \text{cl}(V) \cap B$. Then $\text{cl}(V)$ and $A \cup U \cup B$ are subcontinua of Λ with $\text{cl}(V) \cup (A \cup U \cup B) = \Lambda$ and it follows from Lemma A1(iii) that $\text{cl}(V) \cap \text{cl}(A \cup U \cup B)$ is the union of two disjoint subcontinua A' and B'. It is clear that $A' \cup B'$ separates Λ $(\Lambda - (A' \cup B') = (\Lambda - \text{cl}(V)) \cup (\Lambda - \text{cl}(A \cup U \cup B)))$ and $A' \cup B' \subset A \cup B$. By irreducibility of $\{A, B\}$ we have $A' \cup B' = A \cup B$ and from this it follows that $\text{int}_\Lambda A = \varnothing = \text{int}_\Lambda B$. It follows immediately that $\text{cl}(U) \cap \text{cl}(V) = A \cup B$. □

LEMMA A4. *If P is a proper subcontinuum of the cofrontier Λ with $\text{int}_\Lambda(P) \neq \varnothing$ then $\text{cl}(\Lambda - P) \cap \text{cl}(\Lambda - \text{cl}(\Lambda - P))$ is the union of two disjoint subcontinua A and B and $\{A, B\}$ is an irreducible separating pair.*

PROOF. Since $\Lambda - P$ is connected (Lemma A1(i)) and $\text{int}_\Lambda(P) \neq \varnothing$, $\text{cl}(\Lambda - P)$ is a proper subcontinuum of Λ with nonempty interior (relative to Λ) and $\text{cl}(\Lambda - \text{cl}(\Lambda - P))$ is then also a proper subcontinuum of Λ with nonempty interior relative to Λ. It is clear that

$$\Lambda = \text{cl}(\Lambda - P) \cup \text{cl}(\Lambda - \text{cl}(\Lambda - P))$$

so that, from Lemma A1(iii), we have $\text{cl}(\Lambda - P) \cap \text{cl}(\Lambda - \text{cl}(\Lambda - P)) = A \cup B$ with A and B disjoint subcontinua of Λ. Then $A \cup B$ separates Λ $(\Lambda - (A \cup B) = [\Lambda - \text{cl}(\Lambda - P)] \cup [\Lambda - (\text{cl}(\Lambda - \text{cl}(\Lambda - P)))])$ and there is an irreducible separating pair $\{A', B'\}$ with $A' \subset A$ and $B' \subset B$ by Lemma A2. Let $\Lambda - (A' \cup B') = U \cup V$ with U and V disjoint, connected, and nonempty (Lemma A1(iv)). We have $A' \cup B' = \text{cl}(U) \cap \text{cl}(V)$ (Lemma A3) and, from $A \cup B \supset A' \cup B'$, it follows that $U \supset \Lambda - P$ and $V \supset \Lambda - \text{cl}(\Lambda - P)$ (or vice-versa). Consequently, $\text{cl}(U) \cap \text{cl}(V) \supset \text{cl}(\Lambda - P) \cap \text{cl}(\Lambda - \text{cl}(\Lambda - P)) = A \cup B$. That is, $A' \cup B' \supseteq A \cup B$. Thus $A' = A$, $B' = B$, and $\{A, B\}$ is an irreducible separating pair. □

LEMMA A5. *Suppose that* $\{A, B\}$ *is an irreducible separating pair in the cofrontier* Λ *and let* U *and* V *be the components of* $\Lambda - (A \cup B)$.

  (i) *If* H *is a subcontinuum of* $\mathrm{cl}(U)$ *such that* $H \cap A \neq \varnothing \neq H \cap B$, *then* $H = \mathrm{cl}(U)$.
  (ii) *If* H *is a subcontinuum of* Λ *such that* $H \cap A \neq \varnothing \neq H \cap B$, *then* $H \supset \mathrm{cl}(U)$ *or* $H \supset \mathrm{cl}(V)$.

PROOF. (i) From Lemma A4 we have that $A \cup B = \mathrm{cl}(U) \cap \mathrm{cl}(V)$. In particular, $A \subseteq \mathrm{cl}(V)$ so that $H \cup \mathrm{cl}(V)$ is a subcontinuum of Λ. Also, $H \cap \mathrm{cl}(V) \subseteq A \cup B$ and $H \cap \mathrm{cl}(V)$ meets each of A and B so that $H \cap \mathrm{cl}(V)$ has at least two (and hence exactly two, by Lemma A1) components. It must then be the case (Lemma A1(ii)) that $H \cup \mathrm{cl}(V) = \Lambda$. Thus $H \supset \Lambda - \mathrm{cl}(V) = U$ (Lemma A4) so that $H \supset \mathrm{cl}(U)$. Then $H = \mathrm{cl}(U)$.

To prove (ii), note that $H \cap \mathrm{cl}(U)$ has either one or two components (Lemma A1). If $H \cap \mathrm{cl}(U)$ has two components then $H \cup \mathrm{cl}(U) = \Lambda$ (Lemma A1(ii)). Then $H \supseteq \Lambda - \mathrm{cl}(U) = V$ (by Lemma A4) and $H \supset \mathrm{cl}(V)$. If $H \cap \mathrm{cl}(U)$ is connected then $H \cap \mathrm{cl}(U)$ is a subcontinuum of $\mathrm{cl}(U)$ that meets both A and B so that, by (i), $H \cap \mathrm{cl}(U) = \mathrm{cl}(U)$ and $H \supset \mathrm{cl}(U)$. $\square$

LEMMA A6. *Suppose that* $\{A, B\}$ *and* $\{C, D\}$ *are irreducible separating pairs in the cofrontier* Λ. *Let* U *and* V *be the components of* $\Lambda = (A \cup B)$.

  (i) *If* $A \cap C \neq \varnothing$ *then* $A = C$.
  (ii) *If* $C \cap U \neq \varnothing$ *then* $C \subset U$.
  (iii) *If* $A \cap C = \varnothing$ *then* $\{A, C\}$ *is an irreducible separating pair.*

PROOF. (i) Let $\Lambda - (C \cup D) = U' \cup V'$ with U' and V' connected (Lemma A1(iv)). We have, from Lemma A3, that $A \cup B = \mathrm{cl}(U) \cap \mathrm{cl}(V)$ and $C \cup D = \mathrm{cl}(U') \cap \mathrm{cl}(V')$. Since $\mathrm{cl}(U) \cup \mathrm{cl}(V) = \Lambda$ (Lemma A1(ii)) we have either $\mathrm{cl}(U) \cap D \neq \varnothing$ or $\mathrm{cl}(V) \cap D \neq \varnothing$. Suppose, without loss of generality, that $\mathrm{cl}(U) \cap D \neq \varnothing$. Since $A \subset \mathrm{cl}(U)$ and $C \cap A \neq \varnothing$ we now have that $\mathrm{cl}(U) \cap C \neq \varnothing$. Thus $\mathrm{cl}(U) \supset \mathrm{cl}(U')$ or $\mathrm{cl}(U) \supset \mathrm{cl}(V')$ (Lemma A5(ii)). We may, by proper choice of nomenclature, assume that $\mathrm{cl}(U) \supset \mathrm{cl}(U')$. Then $\Lambda - \mathrm{cl}(U) \subset \Lambda - \mathrm{cl}(U')$ so that $\mathrm{cl}(V) \subset \mathrm{cl}(V')$. Thus $\mathrm{cl}(U) \cup \mathrm{cl}(V') = \Lambda$ and $\mathrm{cl}(U) \cap \mathrm{cl}(V') = H \cup K$ with H and K proper disjoint subcontinua of Λ (Lemma A1(iii)). Now $A \cup C \neq \Lambda$ since $\mathrm{int}_\Lambda A = \varnothing = \mathrm{int}_\Lambda C$ (Lemma A3) so $A \cap C$ is connected (Lemma A1(ii)) and is contained in $\mathrm{cl}(U) \cap \mathrm{cl}(V')$. Thus, exactly one of H and K meets $A \cap C$, say, without loss of generality, $A \cap C \subset H$. Now since $D \subset \mathrm{cl}(U')$ and $\mathrm{cl}(U') \subset \mathrm{cl}(U)$, $D \subset \mathrm{cl}(U)$ and, consequently, $D \subset \mathrm{cl}(U) \cap \mathrm{cl}(V')$. If $D \subset H$ then $H \supset \mathrm{cl}(U')$ or $H \supset \mathrm{cl}(V')$ by Lemma A5(ii). If $H \supset \mathrm{cl}(U')$ then $\mathrm{cl}(U') \subset \mathrm{cl}(V')$, but this is not the case. If $H \supset \mathrm{cl}(V')$ then $K = \varnothing$ and this is not the case. Thus $D \subset K$ so that $\mathrm{cl}(U') \cup K$ is a continuum.

We claim now that either $K \cap B \neq \varnothing$ or $\mathrm{cl}(U') \supset \mathrm{cl}(U)$. To see this, suppose that $B \cap \mathrm{cl}(V') = \varnothing$. Then $B \subset \mathrm{cl}(U')$ so that $\mathrm{cl}(U') \supset \mathrm{cl}(U)$ (by Lemma A5(ii)). If $B \cap \mathrm{cl}(V') \neq \varnothing$ then (since $B \subset \mathrm{cl}(U)$) $B \cap (H \cup K) \neq \varnothing$.

If $B \cap H \neq \varnothing$ then $H \supset \mathrm{cl}(U)$ by Lemma A5(ii). But then $K = \varnothing$ which is not the case. Thus, if $\mathrm{cl}(U')$ does not contain $\mathrm{cl}(U)$, $K \cap B \neq \varnothing$.

We now have that if $\mathrm{cl}(U') \not\supset \mathrm{cl}(U)$ then $\mathrm{cl}(U') \cup K = \mathrm{cl}(U)$ (by Lemma A5(i)) so that $A \subset \mathrm{cl}(U') \cup K$. If $A \cap K \neq \varnothing$ then $K \supset \mathrm{cl}(U)$ by Lemma A5(ii). But then $H = \varnothing$ and this is not the case. Thus $A \cap K = \varnothing$ and $A \subset \mathrm{cl}(U')$.

We have now that, in any case, $A \subset \mathrm{cl}(U')$. We also have that if $\mathrm{cl}(U) \supset \mathrm{cl}(U')$ then $\mathrm{cl}(V) \subset \mathrm{cl}(V')$ and $A \subset \mathrm{cl}(V')$. Then $A \subset \mathrm{cl}(U') \cap \mathrm{cl}(V') = C \cup D$. Since $A \cap C \neq \varnothing$ and C and D are disjoint, $A \subset C$. By a symmetric argument we get $C \subset A$.

For (ii), note that if $C \cap A \neq \varnothing$ or $C \cap B \neq \varnothing$ then $C = A$ or $C = B$ (by (i)), but then $C \cap U = \varnothing$. Thus, if $C \not\subset U$, $C \cap V \neq \varnothing$. Then $U \cup V = U \cup C \cup V$ would be connected and this is not the case.

We now prove (iii). If $A \neq C$ then $A \cap C = \varnothing$ by (i). If $B \cap C \neq \varnothing$ then $C = B$, by (i), so that $\{A, C\} = \{A, B\}$ is an irreducible separating pair. In the same way, if $A \cap D \neq \varnothing$ then $A = D$ and $\{A, C\} = \{C, D\}$ is an irreducible separating pair. Thus we may suppose that $C \subset U \cup V$ and $A \cap D = \varnothing$. Now $C \subset U \cap V$ implies that $C \subset U$ or $C \subset V$ by (ii) and we assume, without loss of generality, that $C \subset U$.

CLAIM. C separates $\mathrm{cl}(U)$.

To establish this claim we consider two cases.

CASE 1. $D \subset V$.

Let U' and V' be the components of $\Lambda - (C \cup D)$. Then $(U' \cap \mathrm{cl}(U)) \cup (V' \cap \mathrm{cl}(U)) = \mathrm{cl}(U) - C$ since $U' \cup V' \cup C = \Lambda - D \supset \mathrm{cl}(U)$. Also, $\mathrm{cl}(U' \cap \mathrm{cl}(U)) \subset \mathrm{cl}(U')$ is disjoint from V', and hence from $V' \cap \mathrm{cl}(U)$, and $\mathrm{cl}(V' \cap \mathrm{cl}(U)) \subset \mathrm{cl}(V')$ is disjoint from U', and hence from $U' \cap \mathrm{cl}(U)$. Furthermore, U is an open set containing C and C is in the frontier of each of U' and V' so that $U' \cap \mathrm{cl}(U) \neq \varnothing \neq V' \cap \mathrm{cl}(U)$. Thus $U' \cap \mathrm{cl}(U)$ and $V' \cap \mathrm{cl}(U)$ are nonempty disjoint open sets (relative to $\mathrm{cl}(U)$) with union $\mathrm{cl}(U) - C$ so that C separates $\mathrm{cl}(U)$.

CASE 2. $D \not\subset V$.

If $D \not\subset V$ then, by (ii), $D \cap V = \varnothing$ and $D \subset \mathrm{cl}(U)$. Again, let U' and V' be the components of $\Lambda - (C \cup D)$. Now C and D are both contained in the continuum $\mathrm{cl}(U)$. By Lemma A5(ii) we have either $\mathrm{cl}(U) \supset \mathrm{cl}(U')$ or $\mathrm{cl}(U) \supset \mathrm{cl}(V')$. Assume, without loss of generality, that $\mathrm{cl}(U) \supset \mathrm{cl}(U')$. By taking complements, then, $V \subset V'$. Thus $\mathrm{cl}(V') \cup \mathrm{cl}(U) \supset \mathrm{cl}(V) \cup \mathrm{cl}(U) = \Lambda$ so that $\mathrm{cl}(V') \cap \mathrm{cl}(U)$ is the union of two disjoint subcontinua H and K (Lemma A1(iii)). Since $\mathrm{cl}(V') \supset \mathrm{cl}(V) \supset A \cup B$, $A \cup B \subset H \cup K$ and each of A and B is contained in one of H or K. We assume $A \subset H$. If B is also contained in H then $H = \mathrm{cl}(U)$ by Lemma A5(i). Then $\mathrm{cl}(V') \supset \mathrm{cl}(U)$ but we also have $\mathrm{cl}(V') \supset V$. Since $\mathrm{cl}(V') \neq \Lambda$, this is not the case. Thus $B \subset K$.

Now C is contained in each of $\mathrm{cl}(V')$ and $\mathrm{cl}(U)$ so $C \subset H \cup K$. Thus

$C \subset H$ or $C \subset K$. If $C \subset H$ (so that $C \cap K = \varnothing$),

$$\mathrm{cl}(U) - C = (H - C) \cup ((\mathrm{cl}(U') - C) \cup K)$$

is a union of two disjoint relatively open (in $\mathrm{cl}(U) - C$) nonempty sets. If $C \subset K$ (so that $C \cap H = \varnothing$), then

$$\mathrm{cl}(U) - C = (K - C) \cup ((\mathrm{cl}(U') - C) \cup H)$$

is a union of two disjoint relatively open (in $\mathrm{cl}(U) - C$) nonempty sets. Thus C separates $\mathrm{cl}(U)$ in Case 2 as well and the claim is established.

Now let L_1 be the component of $\mathrm{cl}(U) - C$ containing A and let L_2 be the component of $\mathrm{cl}(U) - C$ containing B. If $L_1 = L_2$ then $A \cup B \subset \mathrm{cl}(L_1)$ so that $\mathrm{cl}(L_1) = \mathrm{cl}(U)$ (Lemma A5(i)). But then $\mathrm{cl}(U) - C = L_1$ is connected, in contradiction to the previous claim. Thus $L_1 \cap L_2 = \varnothing$.

Now

$$\Lambda - (A \cup C) = (L_1 - A) \cup (L_2 \cup V)$$

expresses $\Lambda - (A \cup C)$ as a disjoint union of relatively open (and hence open in Λ) nonempty sets. Thus $A \cup C$ separates Λ and there is an irreducible separating pair $\{A', C'\}$ with $A' \subset A$ and $C' \subset C$ by Lemma A2. It follows from part (i) of this lemma that $A = A'$ and $C = C'$. Thus $\{A, C\}$ is an irreducible separating pair. $\square$

B. In this section we assemble a few facts about prime ends. All of the statements given without proof or reference are justified in **[M]**. Throughout, Λ will be a cofrontier with complementary domains $\mathscr{D}_e$ and $\mathscr{D}_i$, $\mathscr{D}_e$ being the exterior (unbounded) domain and $\mathscr{D}_i$ the interior (bounded) domain.

If $\mathscr{P}$ is a prime end of $\mathscr{D}_e$ (or of $\mathscr{D}_i$) with representative chain $\{V_n\}$, the *impression of* $\mathscr{P}$ is $I(\mathscr{P}) = \bigcap_{n=1}^{\infty} \mathrm{cl}(V_n)$. $I(\mathscr{P})$ is then a subcontinuum of Λ and is independent of the selected representative chain of $\mathscr{P}$. An open simple arc C in $\mathscr{D}_e$ (or in $\mathscr{D}_i$) is a *cross-cut* if the frontier of C consists of two distinct points in Λ. If $\{V_n\}$ is a prime chain in $\mathscr{D}_e$ (or $\mathscr{D}_i$) defining a prime end $\mathscr{P}$, there is a sequence of cross-cuts $\{C_n\}$ with the properties: $C_n \subset V_n$ for each n; C_n separates $\mathscr{D}_e$ (or $\mathscr{D}_i$) into two domains, one of which, call it U_n, is contained in V_n; $\{U_n\}$ is a prime chain equivalent to $\{V_n\}$. We will say that $\{C_n\}$ is a *representative sequence of cross-cuts* for the prime end $\mathscr{P}$.

Given a sequence of sets X_n in the plane we will denote by $\limsup X_n$ the set of all limit points of sequences $\{x_n\}$ where $x_n \in X_n$ for each n. It follows easily from the definition of the topology on the prime ends that if $\mathscr{P}_n$ is a sequence of prime ends of $\mathscr{D}_e$ (or of $\mathscr{D}_i$) converging to the prime end $\mathscr{P}$ then $\limsup I(\mathscr{P}_n) \subset I(\mathscr{P})$.

Recall that we have denoted (§II) the circles of prime ends in $\hat{\mathscr{D}}_e$ and $\hat{\mathscr{D}}_i$ by $\mathscr{S}_e$ and $\mathscr{S}_i$, respectively.

LEMMA B1. *Suppose that H is a proper subcontinuum of the cofrontier Λ with* $\mathrm{int}_{\Lambda} H \neq \varnothing$. *Let $\hat{H}_e$ ($\hat{H}_i$) be the collection of all prime ends of*

$\mathscr{D}_e$ $(\mathscr{D}_i)$ *that have a representative sequence of cross-cuts* $\{C_n\}$ *with* $\mathscr{F}(C_n) \subset$ H *for all* n. *Then* $\mathrm{cl}(\hat{H}_e)$ $(\mathrm{cl}(\hat{H}_i))$ *is a proper nonempty subarc of* $\mathscr{S}_e$ $(\mathscr{S}_i)$.

PROOF. We will prove the above for $\hat{H}_i$, the proof for $\hat{H}_e$ being the same. To see that $\hat{H}_i \neq \varnothing$, let p be a point of $\mathrm{int}_\Lambda(H)$ accessible from $\mathscr{D}_i$ (points accessible from $\mathscr{D}_i$ are dense in Λ). There is then an $\varepsilon > 0$ small enough so that $\overline{B}_\varepsilon(p) = \{x \in \mathbb{R}^2 \,||\, |x - p| \leq \varepsilon\}$ is disjoint from $\Lambda - H$. Since $\Lambda = \mathscr{F}\mathscr{D}_i$, there is a chain of cross-cuts $\{C_n\}$ in $\mathscr{D}_i$ with $C_n \subset \partial\overline{B}_{\varepsilon/n}(p) \cap \mathscr{D}_i$ for $n = 1, 2, 3 \ldots$ (this follows from [**M**, Theorem 13] since p, being accessible, is a principal point). Then $\mathscr{F}C_n \subset H$ for all $n \geq 1$ so that $\{C_n\}$ defines a prime end in $\hat{H}_i$.

To prove that $\hat{H}_i$ is connected, let $\{C_n\}$ and $\{C'_n\}$ be sequences of cross-cuts defining the distinct elements $\mathscr{P}$ and $\mathscr{P}'$ of $\hat{H}_i$. By discarding a finite number of initial elements of these sequences we may assume that $\mathscr{F}C_1 \cup \mathscr{F}C'_1 \subset H$ and $\mathrm{cl}(C_1) \cap \mathrm{cl}(C'_1) = \varnothing$. Now let C be a simple arc in $\mathscr{D}_i$ with endpoints x and x' and with $C \cap C_1 = \{x\}$, $C \cap C'_1 = \{x'\}$. Let α and α' be half-open arcs in C_1 and C'_1 with endpoints x and x' and frontier in Λ (see Figure 3). Then $\gamma = \alpha \cup C \cup \alpha'$ is a cross-cut in $\mathscr{D}_i$ that separates $\mathscr{D}_i$ into two domains, one of which, call it U, has frontier contained in $\gamma \cup H$. By replacing α by $\mathscr{D}_i \cap (\mathrm{cl}(C_1 - \alpha))$ or α' by $\mathscr{D}_i \cap (\mathrm{cl}(C'_1 - \alpha'))$ if necessary, we may arrange that U_1 and U'_1 are contained in U (U_1 and U'_1 being the domains of $\mathscr{D}_i - C_1$ and $\mathscr{D}_i - C'_1$ containing C_2 and C'_2, respectively). Now let $\hat{L}$ be the collection of all prime ends of $\mathscr{D}_i$ that have defining chains all of whose members lie in U. Then $\hat{L}$ is an (open) arc in $\mathscr{S}_i$, containing $\mathscr{P}$ and $\mathscr{P}'$, the members of which all lie in $\hat{H}_i$. Thus $\hat{H}_i$ is connected.

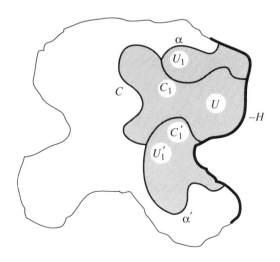

FIGURE 3

To show that $\mathrm{cl}(\hat{H}_i)$ is proper in $\mathscr{S}_i$ it suffices to show that there are at least two distinct prime ends in the complement of $\hat{H}_i$. Let p_1 and p_2 be distinct accessible points of $\Lambda - H$ and let α_1 and α_2 be arcs in $\mathscr{D}_i$ accessing p_1 and p_2. Then α_1 and α_2 represent distinct prime ends $\mathscr{P}_1$ and $\mathscr{P}_2$ whose impressions are not contained in H because they contain p_1 and p_2, respectively (see [**M**, Theorem 17.1]). The prime ends $\mathscr{P}_1$ and $\mathscr{P}_2$ therefore do not belong to $\hat{H}_i$. □

Suppose that G is an orientation-preserving homeomorphism of a circle S. According to Denjoy [**D**], a necessary and sufficient condition for G to have a periodic point of period $n \geq 2$ in S is that the rotation number of G be an irreducible fraction of the form m/n; a necessary and sufficient condition for the existence of a fixed point is that the rotation number be zero.

As described in §II, an orientation-preserving homeomorphism F of the plane with invariant cofrontier Λ induces orientation-preserving homeomorphisms $\hat{F}_i$ and $\hat{F}_e$ on the circles $\mathscr{S}_i$ and $\mathscr{S}_e$ of prime ends associated with Λ. We denote the rotation numbers of these circle homeomorphisms by ρ_i and ρ_e.

LEMMA B2. *Let F, Λ, ρ_i, and ρ_e be as above and suppose that H and K are proper subcontinua of Λ with $\Lambda = H \cup K$.*

(i) *If $F(H) = H$ then $\rho_e = 0 = \rho_i$.*

(ii) *If $F(H) = K$ and $F(K) = H$ then $\rho_e = 1/2 = \rho_i$.*

PROOF. We prove (i) and (ii), for ρ_i. The proofs for ρ_e are obtained by replacing all i's by e's. Since H and K are proper in Λ and $H \cup K = \Lambda$, $\mathrm{int}_\Lambda(H) \neq \varnothing \neq \mathrm{int}_\Lambda(K)$. Let $\hat{H}_i$ and $\hat{K}_i$ be as in Lemma B1. If $\{C_n\}$ is a sequence of cross-cuts representing the prime end $\mathscr{P}$ of $\mathscr{D}_i$ then $\{F(C_n)\}$ is a sequence of cross-cuts representing the prime end $\hat{F}_i(\mathscr{P})$. Supposing that $F(H) = H$ it follows that $\hat{F}_i(\hat{H}_i) = \hat{H}_i$. The proper closed arc $\mathrm{cl}(\hat{H}_i)$ in $\mathscr{S}_i$ is then invariant under $\hat{F}_i$ and must contain a fixed point. Thus $\rho_i = 0$ and (i) is established.

If $F(H) = K$ and $F(K) = H$ then $\hat{F}_i(\hat{H}_i) = \hat{K}_i$ and $\hat{F}_i(\hat{K}_i) = \hat{H}_i$. Now, the element of $\hat{H}_i$ constructed in the proof of Lemma B1 to show that $\hat{H}_i \neq \varnothing$, had a point $p \in \mathrm{int}_\Lambda(H)$ in its impression. If the construction is carried out with $p \in H - K$, we obtain a prime end of $\hat{H}_i$ with impression not contained in K. Since every element of $\mathrm{cl}(\hat{K}_i)$ has impression contained in K, we see that $\mathrm{cl}(\hat{H}_i) \neq \mathrm{cl}(\hat{K}_i)$. Now since $\hat{F}_i^2(\hat{H}_i) = \hat{H}_i$, $\hat{F}_i^2$ has a fixed point in $\mathrm{cl}(\hat{H}_i)$. Since $\hat{F}_i$ is an orientation-preserving homeomorphism of the circle $\mathscr{S}_i$ and $\hat{F}_i(\hat{H}_i) = \hat{K}_i \neq \hat{H}_i$, this fixed point of $\hat{F}_i^2$ must be a periodic point of period 2. Thus $\rho_i = 1, 2$. □

LEMMA B3 (Cartwright and Littlewood). *Suppose that F is an orientation-preserving homeomorphism of the plane with invariant cofrontier Λ. If $\mathscr{P}$ is a prime end of $\mathscr{D}_i$ (or of $\mathscr{D}_e$) the impression of which contains a fixed* *point of* F *and* $\rho_i \neq 0$ $(\rho_e \neq 0)$, *then every fixed point of* $F|_\Lambda$ *is in the impression* $I(\mathscr{P})$ *of* $\mathscr{P}$. *Furthermore, there is a positive integer* M *such that* $I(\mathscr{P}) \cup F^M(I(\mathscr{P})) = \Lambda$.

PROOF. That all fixed points of $F|_\Lambda$ must belong to $I(\mathscr{P})$ is part of the conclusion of Corollary 5 following Theorem 6 of [CL2]. This also clearly follows from $I(\mathscr{P}) \cup F^M(I(\mathscr{P})) = \Lambda$ for some $M \geq 1$. This last result is easily culled from the proof of Theorem 7 in [CL2]. The idea is as follows. Suppose that it is ρ_e that is nonzero and let p be a fixed point of F in Λ. Let $\mathscr{P}$ be a prime end of $\mathscr{D}_e$ with $p \in I(\mathscr{P})$ (there is such a $\mathscr{P}$ since $\mathscr{F}D_e = \Lambda$). One can construct a simple half-open arc C in $\mathscr{D}_e$ such that $\mathscr{F}C = I(\mathscr{P})$ and $F(C) \cap C = \varnothing$ (this last because $\rho_e \neq 0$). If $\mathscr{D}_e$ is mapped one-to-one and conformally onto $|z| > 1$ in the complex plane by ϕ (with points near Λ going to points near $|z| = 1$) then the homeomorphism $\phi \circ F \circ \phi^{-1}$ extends to a homeomorphism $\tilde{F}$ of $|z| \geq 1$ that, on the unit circle, is topologically conjugate with $\hat{F}_e$ on $\mathscr{S}_e$. Thus the rotation number of $\hat{F}|_{|z|=1}$ equals ρ_e and so is nonzero.

Now in $\mathscr{D}_e$ one can join the endponts of the half-open arcs C and $F(C)$ lying in $\mathscr{D}_e$ by a simple arc to obtain an (open) arc C', the frontier of which lies in Λ. Since the fixed point p is in $I(\mathscr{P}) = \mathscr{F}C$, p is also in $F(I(\mathscr{P})) = I(\hat{F}_e(\mathscr{P})) = \mathscr{F}(F(C)) = F(\mathscr{F}(C))$. That is, $\mathscr{F}(F(C)) \cap \mathscr{F}C \neq \varnothing$, so that $C' \cup \mathscr{F}C \cup \mathscr{F}(F(C))$ separates the plane. If $\mathscr{F}C \cup \mathscr{F}(F(C)) \neq \Lambda$, this separation is into two domains; otherwise we are done with $M = 1$. Now if $\mathscr{F}C \cup \mathscr{F}(F(C)) \neq \Lambda$, the arc connecting C with $F(C)$ to form C' can be taken in such a way that the bounded complementary domain U of $C' \cup \mathscr{F}C \cup \mathscr{F}(F(C))$ is contained in $\mathscr{D}_e$ (there are essentially two ways of connecting C and $F(C)$—these go around Λ in opposite directions; see Figure 4). It is automatic that $\mathscr{F}U \cap \Lambda \subset I(\mathscr{P}) \cup F(I(\mathscr{P}))$. Now, under ϕ, C' is taken to an open arc $\phi(C')$ with frontier $\mathscr{F}(\phi(C')) = \{z_0, z_1\}$ contained in the unit circle and with $\tilde{F}(z_0) = z_1$.

From the nonzero rotation number of $\tilde{F}|_{|z|=1}$ it follows that, for some

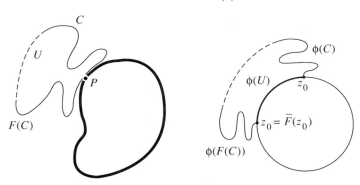

FIGURE 4

$N \geq 1$, $\bigcup_{n=0}^{N} \tilde{F}^{n}([z_0, z_1])$ is the entire unit circle where $[z_0, z_1]$ is the arc in $|z| = 1$, contained in $\mathscr{F}(\phi(U))$, from z_0 to z_1. Transferring back to $\mathscr{D}_e$ via ϕ^{-1}, one finds that $\bigcup_{n=0}^{N} F^{n}(\mathscr{F}(C)) = \Lambda$.

The fixed point p is in all of the subcontinua $F^{n}(\mathscr{F}(C))$ of Λ. Lemma 41 of [CL2], which states that if H_1, H_2, and H_3 are subcontinua of the cofrontier Λ such that $H_1 \cup H_2 \cup H_3 = \Lambda$ and $H_1 \cap H_2 \cap H_3 \neq \varnothing$ then one of H_1, H_2, and H_3 is contained in the union of the other two, can now be applied to reduce $\bigcup_{n=0}^{N} F^{n}(\mathscr{F}(C)) = \Lambda$ to $\mathscr{F}(C) \cup F^{M}(\mathscr{F}(C)) = \Lambda$ for some M, $1 \leq M \leq N$. That is, $I(\mathscr{P}) \cup F^{M}(I(\mathscr{P})) = \Lambda$. □

C. The properties of the Alexander-Spanier cohomology groups that we use in this section can be found in Spanier [S]. All homology and cohomology groups will be taken with integer ($\mathbb{Z}$) coefficients.

Suppose that Λ is a cofrontier in the plane and suppose that $\Lambda = P \cup \mathscr{Q}$, where P and $\mathscr{Q}$ are proper subcontinua of Λ. Under these circumstances, $P \cap \mathscr{Q}$ has exactly two components A and B (Lemma A1), and none of the continua $P, \mathscr{Q}, A$ or B separates the plane. According to the Alexander duality theorem,

$$\check{H}^{1}(\Lambda) \cong H_{1}(\mathbb{R}^{2}, \mathbb{R}^{2} \backslash \Lambda) \cong \tilde{H}_{0}(\mathbb{R}^{2} \backslash \Lambda) \cong \mathbb{Z},$$

where $\check{H}^{1}$ denotes Alexander-Spanier cohomology, and $\tilde{H}_{0}$ and H_{1} denote singular homology, $\tilde{H}_{0}$ being the reduced group. Alexander duality also yields

$$\check{H}^{1}(P) \cong \check{H}^{1}(\mathscr{Q}) \cong \check{H}^{1}(A) \cong \check{H}^{1}(B) \cong 0.$$

Suppose now that a generator γ of $\check{H}^{1}(\Lambda)$ has been selected. The selection of γ may be thought of as giving a sense of direction to the subcontinua of Λ. The goal here is to use γ to determine unambiguously which of the components of $P \cap \mathscr{Q}$ is at the beginning of P and the terminus of $\mathscr{Q}$, and which is at the terminus of P and the beginning of $\mathscr{Q}$. (See Figure 5.)

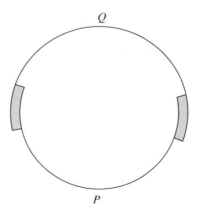

FIGURE 5

It will be seen that this may be accomplished by considering the relative 1-cocycle of P modulo $P \cap \mathscr{Q}$ which corresponds to the generator γ, and then determining a corresponding 0-cocycle for $P \cap \mathscr{Q}$ by way of the Mayer-Vietoris sequence for the pair $(P, \mathscr{Q})$.

The pair $(P, \mathscr{Q})$ is excisive for Alexander-Spanier cohomology, and the associated Mayer-Vietoris sequence contains the following segment:

$$\tilde{H}^0(P) \oplus \tilde{H}^0(\mathscr{Q}) \to \tilde{H}^0(P \cap \mathscr{Q}) \to \check{H}^1(\Lambda) \to \check{H}^1(P) \oplus \check{H}^1(\mathscr{Q}).$$

Here the symbol "$\check{}$" has been suppressed on the reduced groups. Since $\tilde{H}^0$ and $\check{H}^1$ vanish for P and $\mathscr{Q}$, we have the isomorphism

$$\tilde{H}^0(P \cap \mathscr{Q}) \overset{\to}{\cong} \check{H}^1(\Lambda).$$

This isomorphism is the composite of three isomorphisms

(1) $\tilde{H}^0(P \cap \mathscr{Q}) \to \check{H}^1(P, P \cap \mathscr{Q}) \overset{\text{excision}}{\longleftarrow} \check{H}^1(\Lambda, \mathscr{Q}) \to \check{H}^1(\Lambda),$

and by the hexagonal lemma [Gr] it is also the negative of the composite of the following three isomorphisms:

(2) $\tilde{H}^0(P \cap \mathscr{Q}) \to \check{H}^1(\mathscr{Q}, P \cap \mathscr{Q}) \overset{\text{excision}}{\longleftarrow} \check{H}^1(\Lambda, P) \to \check{H}^1(\Lambda).$

The reduced group $\tilde{H}^0(P \cap \mathscr{Q})$ may be identified with the group of locally constant integer-valued functions on $P \cap \mathscr{Q}$ modulo the subgroup of constant functions. Therefore the generators of $\tilde{H}^0(P \cap \mathscr{Q})$ may be taken to be the characteristic functions of the two components of $P \cap \mathscr{Q}$. Suppose that the Mayer-Vietoris isomorphism (1) for the pair $(P, \mathscr{Q})$ carries the selected generator γ of $\check{H}^1(\Lambda)$ onto the generator of $\tilde{H}^0(P \cap \mathscr{Q})$ represented by $p: P \cap \mathscr{Q} \to \{0, 1\}$. Then $A = p^{-1}(0)$ and $B = p^{-1}(1)$ are the two components of $P \cap \mathscr{Q}$ and A will be said to begin P and terminate $\mathscr{Q}$ and B will be said to terminate P and begin $\mathscr{Q}$.

If all is repeated for the reversed pair $(\mathscr{Q}, P)$, then the Mayer-Vietoris isomorphism is given by (2), which is the negative of (1), and therefore γ is taken onto the generator of $\tilde{H}^0(P \cap \mathscr{Q})$ represented by $q: P \cap \mathscr{Q} \to \{0, 1\}$ with $q^{-1}(0) = p^{-1}(1) = B$ and $q^{-1}(1) = p^{-1}(0) = A$. That is, B begins $\mathscr{Q}$ and terminates P and A terminates $\mathscr{Q}$ and begins P, as before.

The situation is summarized by the following definition in which the notation $(\mathscr{Q} : P)_\gamma$, or simply $\mathscr{Q} : P$, is to be thought of as denoting the component of $P \cap \mathscr{Q}$ which terminates $\mathscr{Q}$ and begins P, while $P : \mathscr{Q}$ is the component which terminates P and begins $\mathscr{Q}$.

DEFINITION. If γ is a generator of $\check{H}^1(\Lambda)$ and if $p: P \cap \mathscr{Q} \to \{0, 1\}$ represents the reduced 0-cocycle on $P \cap \mathscr{Q}$ which corresponds to γ under (1), then $(\mathscr{Q} : P)_\gamma = p^{-1}(0)$ and $(P : \mathscr{Q})_\gamma = p^{-1}(1)$.

We proceed to develop properties of this construction.

LEMMA C1. *Suppose that* P_1, P_2, *and* $\mathcal{Q}$ *are proper subcontinua of the cofrontier* Λ *such that* $P_1 \cup \mathcal{Q} = P_2 \cup \mathcal{Q} = \Lambda$. *If* $P_1 \subset P_2$, *then* $\mathcal{Q} : P_1 \subset \mathcal{Q} : P_2$ *and* $P_1 : \mathcal{Q} \subset P_2 : \mathcal{Q}$.

PROOF. The Mayer-Vietoris isomorphisms given by (1) for the pairs $(P_1, \mathcal{Q})$ and $(P_2, \mathcal{Q})$ form the rows in the following commuting diagram in which the vertical arrows denote homomorphisms induced by inclusions:

$$\begin{array}{ccccccc}
\check{H}^0(P_2 \cap \mathcal{Q}) & \to & \check{H}^1(P_2, P_2 \cap \mathcal{Q}) & \leftarrow & \check{H}^1(\Lambda, \mathcal{Q}) & \to & \check{H}^1(\Lambda) \\
\downarrow & & \downarrow & & \| & & \| \\
\check{H}^0(P_1 \cap \mathcal{Q}) & \to & \check{H}^1(P_1, P_1 \cap \mathcal{Q}) & \leftarrow & \check{H}^1(\Lambda, \mathcal{Q}) & \to & \check{H}^1(\Lambda).
\end{array}$$

Since the diagram commutes, the 0-cocycle representative $p_2: P_2 \cap \mathcal{Q} \to \{0, 1\}$ corresponding to γ under the top row of (3) restricts to the 0-cocycle representative $p_1: P_1 \cap \mathcal{Q} \to \{0, 1\}$ which corresponds to γ under the bottom row of (3). Consequently, $\mathcal{Q}: P_1 = p_1^{-1}(0) \subset p_2^{-1}(0) = \mathcal{Q}: P_2$, and $P_1: \mathcal{Q} = p_1^{-1}(1) \subset p_2^{-1}(1) = P_2: \mathcal{Q}$. $\square$

Suppose now that $\{A, B\}$ is an irreducible separating pair in the cofrontier Λ (§III(A)). If P and $\mathcal{Q}$ denote the closures of the components of $\Lambda - (A \cup B)$, then $\Lambda = P \cup \mathcal{Q}$, and A and B are the components of $P \cap \mathcal{Q}$ (Lemma A3). If $(\mathcal{Q}: P)_\gamma = A$, then $(P: \mathcal{Q})_\gamma = B$ and the notation $[A, B]_\gamma$ (or just $[A, B]$) will be used for P. In that case $\mathcal{Q}$ will be denoted by $[B, A]_\gamma$ (or $[B, A]$). In keeping with interval notation, $[A, B] - (A \cup B)$ will be denoted by (A, B), $[A, B] - A$ by $(A, B]$, and $[A, B] - B$ by $[A, B)$. The notation will be extended to the case in which one member of a separating pair is repeated by setting $[A, A] = A$.

LEMMA C2. *Suppose that* $\{A, B\}$ *and* $\{C, D\}$ *are irreducible separating pairs in the cofrontier* Λ. *Then:*

(i) $[A, B] : [B, A] = B$, $[B, A] : [A, B] = A$, *and* $(A, B) = \Lambda - [B, A]$;

(ii) $[A, B] = \mathrm{cl}(A, B)$, (A, B) *is open in* Λ, *and* (A, B) *and* (B, A) *are the components of* $\Lambda - (A \cup B)$;

(iii) $A \cup B = \mathscr{F}_\Lambda(A, B)$ *and* $(A, B) = \mathrm{int}_\Lambda[A, B]$;

(iv) *if* $C \cap [A, B] \neq \varnothing$, *then* $C \subset [A, B]$;

(v) *if* $C \cap (A, B) \neq \varnothing$, *then* $C \subset (A, B)$; *and*

(vi) *if* $[A, B] = [C, D]$, *then* $A = C$ *and* $B = D$.

PROOF. According to Lemma A3, $[A, B] \cap [B, A] = A \cup B$ and therefore, $(A, B) = \Lambda - [B, A]$. Letting $P = [A, B]$ and $\mathcal{Q} = [B, A]$, it follows from the definition of the notation that $\mathcal{Q}: P = A$ and $P: \mathcal{Q} = B$. This proves (i) and also shows that (A, B) is a connected open subset of Λ (Lemma A1). Therefore, the components of $\Lambda - (A \cup B)$ are (A, B) and (B, A), and by definition, $[A, B]$ and $[B, A]$ are the closures of these components; this proves (ii). Item (iii) is a consequence of (ii) since $[A, B] - (A, B) = A \cup B$.

Since $[A, B] = (A, B) \cup A \cup B$, (iv) and (v) follow from (ii) and Lemma A6.

To prove (vi), observe that, by (iii), $(A, B) = \text{int}_\Lambda[A, B] = \text{int}_\Lambda[C, D] = (C, D)$ and therefore $A \cup B = \mathcal{F}(A, B) = \mathcal{F}(C, D) = C \cup D$. There are, since A and B are each connected and they are disjoint and the same is true of C and D, only two possibilities: $A = C$ and $B = D$, or $A = D$ and $B = C$. But $[A, B] \neq [B, A]$ so the second possibility is ruled out. □

LEMMA C3. *Suppose that* $\{A, B\}$ *and* $\{C, D\}$ *are irreducible separating pairs in the cofrontier* Λ, *and suppose that* F *is an orientation-preserving homeomorphism of* $\mathbb{R}^2$ *such that* $F(\Lambda) = \Lambda$.

   (i) *If* $C \subset [A, B]$, *then* $[A, B] = [A, C] \cup [C, B]$ *and* $[A, C] \cap [C, B] = C$.

   (ii) *If* $[A, B] \cup [C, D] = \Lambda$, *then* $[A, B] \cap [C, D] = [C, B] \cup [A, D]$.

   (iii) $F([A, B]]) = [F(A), F(B)]$.

PROOF. If C intersects $A \cup B$, then $C = A$ or $C = B$ (Lemma A6). If $C = A$ or $C = B$ then (i) is obvious. So in proving (i) it may be assumed that $C \subset (A, B)$, in which case, by Lemma A6, $\{A, C\}$ and $\{C, B\}$ are also irreducible separating pairs. Since $C \subset (A, B)$ and $(A, B) \cap [B, A] = \varnothing$ (Lemma C2), the connected set (B, A) is disjoint from $A \cup C$, and hence is contained in one of the components (A, C) or (C, A) of $\Lambda - (A \cup C)$. It will be shown that $(B, A) \subset (A, C)$.

Suppose, for the purpose of contradiction, that $(B, A) \subset (A, C)$. The result of taking closures and complements in Λ is the pair of containments:

$$[B, A] \subset [A, C] \quad \text{and} \quad [C, A] \subset [A, B].$$

Using these in connection with Lemma C1 results in

$$A = [C, A] : [A, C] \subset [A, B] : [A, C] \supset [A, B] : [B, A] = B$$

which shows that the subcontinuum $[A, B] : [A, C]$ of $[A, B]$ contains both A and B and hence is equal to $[A, B]$ (Lemma A5). Since $[C, A] \subset [A, B]$, this leads to a contradiction:

$$[C, A] \subset [A, B] = [A, B] : [A, C] \subset [A, C].$$

Consequently, $(B, A) \subset (C, A)$, as desired, and by taking closures and complements in Λ,

$$[B, A] \subset [C, A] \quad \text{and} \quad [A, C] \subset [A, B].$$

By repeating the argument for the separation $\Lambda - (C \cup B) = (B, C) \cup (C, B)$, it can be shown that $(B, A) \subset (B, C)$ and hence that $[B, A] \subset [B, C]$ and $[C, B] \subset [A, B]$. Thus from the hypothesis that $C \subset (A, B)$ it follows that $[A, C] \cup [C, B] \subset [A, B]$.

To see that this last containment is an equality observe that since $[A, C] \cup [C, B] \neq \Lambda$ and $[A, C] \cap [C, B] \supset C$, $[A, C] \cup [C, B]$ is a subcontinuum of $[A, B]$ containing both A and B. By Lemma A5, it must be all of $[A, B]$.

To complete the proof of (i) it remains to be shown that $[A, C] \cap [C, B] = C$. We continue to assume that $C \cap (A \cup B) = \varnothing$ as otherwise this is obvious.

Now if $B \cap [A, C] \neq \varnothing$ then $B \subset [A, C]$ (Lemma A6) so that, from what has been established above, $[A, C] = [A, B] \cup [B, C]$, and in particular $[A, B] \subset [A, C]$. Since $[A, C] \subset [A, B]$ we have $[A, B] = [A, C]$ and by Lemma C2, $B = C$ contrary to our assumption. Thus $B \subset \Lambda - [A, C] = (C, A)$. From the last containment it follows that $[C, A] = [C, B] \cup [B, A]$ and now

$$[A, C] \cap [C, B] \subset [A, C] \cap [C, A] = A \cup C.$$

It can be argued that $A \cap [C, B] = \varnothing$ just as it was argued above that $B \cap [A, C] = \varnothing$. Thus $[A, C] \cap [C, B] \subset C$ and the proof of (i) is complete.

To prove (ii), suppose that $[A, B] \cup [C, D] = \Lambda$. Then (Lemma C2) $[B, A] \subset [C, D]$ and $[D, C] \subset [A, B]$. In particular, by part (i) of this lemma, $[A, B] = [A, C] \cup [C, B]$ and $[A, C] \cap [C, B] = C$, so that $D \subset [A, C]$ or $D \subset [C, B]$. A contradiction results from supposing that $D \subset [C, B]$ because then (again by (i)) $[A, B] = [A, C] \cup [C, D] \cup [D, B] \supset [C, D] \supset [B, A]$, which is impossible. Therefore it must be that $D \subset [A, C]$ and it follows that

$$[A, B] = [A, D] \cup [D, C] \cup [C, B].$$

Notice that this equation is valid in case $A = D$ or $B = C$, and it follows from this equation that $A \neq C$ and $B \neq D$. More generally, $[A, D] \cap [C, B] = \varnothing$ since $[A, D] \subset [A, C]$ and $[A, C] \cap [C, B] = C$, while $[C, B] \subset [D, B]$ and $[A, D] \cap [D, B] = D$, so that $[A, D] \cap [C, B] \subset C \cap D = \varnothing$.

In the same way it can be shown that

$$[C, D] = [C, B] \cup [B, A] \cup [A, D].$$

The intersection of $[A, B]$ with $[C, D]$ can now be calculated as the union of nine intersections including $[A, D] \cap [B, A] \subset [A, B] \cap [B, A] = A \cup B$ and three more of the same sort. The end result is that $[A, B] \cap [C, D] = [A, D] \cup [B, C]$.

For the proof of (iii), first note that $\{F(A), F(B)\}$ is also an irreducible separating pair as this is a topological property. In fact the decomposition $\Lambda = [A, B] \cup [B, A]$, $[A, B] \cap [B, A] = A \cup B$, is carried by F to the decomposition $\Lambda = F([A, B]) \cup F([B, A])$, $F([A, B]) \cap F([B, A]) = F(A) \cup F(B)$. Thus there are two possibilities: $F([A, B]) = [F(A), F(B)]$ or $F([A, B]) = [F(B), F(A)]$. What must be shown is that $F([B, A]): F([A, B]) = F(A)$.

The Mayer-Vietoris isomorphism (1) may be rewritten for $[A, B]$ in the form

(4) $\qquad \check{H}^0(A \cup B) \to \check{H}^1([A, B], A \cup B) \leftarrow \check{H}^1(\Lambda, [B, A]) \to \check{H}^1(\Lambda).$

The image of the generator γ of $\check{H}^1(\Lambda)$ in $\check{H}^0(A \cup B)$ is represented by the function (0-cocycle) $p: A \cup B \to \{0, 1\}$ given by $P(A) = 0$ and $P(B) = 1$. Since all the isomorphisms in (1) are functional, there is a commuting diagram obtained by applying F to (4), in which all vertical arrows represent isomorphisms induced by F:

(5)

$$\check{H}^0(A \cup B) \quad \to \quad \check{H}^1([A, B], A \cup B) \quad \leftarrow \quad \check{H}^1(\Lambda, [B, A]) \quad \to \quad \check{H}^1(\Lambda)$$
$$\uparrow \qquad\qquad \uparrow \qquad\qquad \uparrow \qquad\qquad \uparrow$$
$$\check{H}^0(F(A) \cup F(B)) \to \check{H}^1(F([A, B]), F(A) \cup F(B)) \leftarrow \check{H}^1(\Lambda, F([B, A])) \to \check{H}^1(\Lambda).$$

What will be shown below is that, in the right-most vertical isomorphism of (5), $F^*(\gamma) = \gamma$. It then follows that, under the isomorphism in the bottom row of (5), γ is taken to the generator of $\check{H}^0(F(A) \cup F(B))$ represented by $q: F(A) \cup F(B) \to \{0, 1\}$ given by $q(F(A)) = 0$, $q(F(B)) = 1$. Then, by definition, $F([B, A]): F([A, B]) = F(A)$, as required.

To show that $F^*(\gamma) = \gamma$, the orientation of the plane must be related to the choice of generator γ in $\check{H}^1(\Lambda)$. Let $\mathscr{D}_i$ and $\mathscr{D}_e$ be the complementary domains of Λ in $\mathbb{R}^2$ and let S^2 be the one-point compactification of $\mathbb{R}^2$. Further, let $\overline{\mathscr{D}}_i$ and $\overline{\mathscr{D}}_e$ be the closures of $\mathscr{D}_i$ and $\mathscr{D}_e$ in S^2 (so $\overline{\mathscr{D}}_i = \mathscr{D}_i \cup \Lambda$ and $\overline{\mathscr{D}}_e = \mathscr{D}_e \cup \Lambda \cup \{\infty\}$). We have: $\overline{\mathscr{D}}_i \cup \overline{\mathscr{D}}_e = S^2$; $\overline{\mathscr{D}}_i \cap \overline{\mathscr{D}}_e = \Lambda$; $S^2 - \overline{\mathscr{D}}_e = \mathscr{D}_i$; and $S^2 - \overline{\mathscr{D}}_i = \mathscr{D}_e$.

Now F extends to a homeomorphism $\overline{F}$ of S^2 that is necessarily of degree 1 on S^2, in the sense that $\overline{F}^*$ is the identity on $\check{H}^2(S^2) \cong \mathbb{Z}$, since F is orientation-preserving on $\mathbb{R}^2$. Also, $\overline{\mathscr{D}}_i$ and $\overline{\mathscr{D}}_e$ are invariant under $\overline{F}$. The functorial nature of the Mayer-Vietoris sequence for the pair $(\overline{\mathscr{D}}_i, \overline{\mathscr{D}}_e)$ then leads to the commuting diagram with exact rows:

(6)
$$\check{H}^1(\overline{\mathscr{D}}_i) \oplus \check{H}^1(\overline{\mathscr{D}}_e) \quad \to \quad \check{H}^1(\Lambda) \quad \to \quad \check{H}^2(S^2)$$
$$\uparrow \qquad\qquad \uparrow F^* \qquad \uparrow \overline{F}^* = \mathrm{id}$$
$$\check{H}^1(\overline{\mathscr{D}}_i) \oplus \check{H}^1(\overline{\mathscr{D}}_e) \quad \to \quad \check{H}^1(\Lambda) \quad \to \quad \check{H}^2(S^2).$$

Now $\overline{\mathscr{D}}_i$ and $\overline{\mathscr{D}}_e$, viewed as planar continua, do not separate the plane. It follows from Alexander duality that $\check{H}^1(\overline{\mathscr{D}}_i) \cong \check{H}^1(\overline{\mathscr{D}}_e) \cong 0$. The homeomorphism $\check{H}^1(\Lambda) \to \check{H}^2(S^2)$ in (6) is thus an injection and it follows (from commutativity of (6)) that $F^*(\gamma) = \gamma$ as desired. □

D. In this section we give the proofs of the theorems and their corollaries. We begin with a proposition.

PROPOSITION D1. *Suppose that F is an orientation-preserving homeomorphism of the plane with invariant cofrontier Λ. If $\Lambda = H \cup K$ with H and K proper indecomposable subcontinua of Λ then either*

(i) *$F(H) = H$, $F(K) = K$, F has a fixed point in Λ, and the prime end rotation numbers, ρ_i and ρ_e, of F associated with Λ are both zero; or*

(ii) *$F(H) = K$, $F(K) = H$, all periodic points of F in Λ have even period, and $\rho_i = 1/2 = \rho_e$.*

PROOF. Let $\Lambda = H \cup K$ with H and K proper indecomposable subcontinua of Λ. Then $\mathrm{int}_\Lambda(H) \neq \varnothing \neq \mathrm{int}_\Lambda(K)$. In the following we use the fact (see [**HY**, Theorem 3-41]) that a continuum is indecomposable if and only if it has no proper subcontinua with nonempty interior.

According to Lemma A1(iii), $H \cap K$ is the union of two disjoint continua A and B. Since A and B are proper subcontinua of H and H is indecomposable, $\text{int}_H(A) = \varnothing = \text{int}_H(B)$. Similarly $\text{int}_K(A) = \varnothing = \text{int}_K(B)$. It follows that $H = \text{cl}(\Lambda - K)$ and $K = \text{cl}(\Lambda - H)$ and that $\{A, B\}$ is an irreducible separating pair in Λ (Lemma A4).

Suppose that $F(H) \cup H = \Lambda$. Then $F(H) \supset \Lambda - H$ so that $F(H) \supset \text{cl}(\Lambda - H) = K$. Since $F(H)$ is also indecomposable and $\text{int}_{F(H)}(K) \neq \varnothing$, $F(H) = K$. Then $F(K) \supset \Lambda - K$ and, as above, $F(K) = H$.

Suppose that $F(H) \cup K = \Lambda$. Then $F(H) \supset \Lambda - K$ so that $F(H) \supset H$ and, arguing as above, $F(H) = H$. Then $F(K) \supset \Lambda - H$ so that $F(K) \supset K$ and $F(K) = K$.

Suppose now that $F(H) \cup H \neq \Lambda \neq F(H) \cup K$. Then $F(H) \cap H$ and $F(H) \cap K$ are subcontinua of $F(H)$ by Lemma A1(ii). Now $F(H) = (F(H) \cap H) \cup (F(H) \cap K)$ and $F(H)$ is indecomposable, so either $F(H) \cap H = F(H)$ or $F(H) \cap K = F(H)$. That is, either $F(H) \subset H$ or $F(H) \subset K$. Since H and K are indecomposable and $\text{int}_H(F(H)) \neq \varnothing \neq \text{int}_K(F(H))$, we have that either $F(H) = H$ or $F(H) = K$. As above, if $F(H) = H$ then $F(K) = K$ and if $F(H) = K$ then $F(K) = H$.

Thus, in any case, we have that either $F(H) = H$ and $F(K) = K$ or $F(H) = K$ and $F(K) = H$.

The conclusions on ρ_i and ρ_e in parts (i) and (ii) of this proposition follow from Lemma B2.

From the definition of a cofrontier we have that, H being proper in Λ, H does not separate the plane. Thus if $F(H) = H$, F has a fixed point in H by a theorem of Cartwright and Littlewood (Theorem 5 of [**CL2**]) and we have the situation of (i).

Now suppose that $F(H) = K$ and $F(K) = H$. Let $H \cap K = A \cup B$ with, as previously established, $\{A, B\}$ an irreducible separating pair in Λ. Let $[A, B]$ be as defined in §III(C). Then $[A, B] = H$ or $[A, B] = K$ and we may as well assume that $[A, B] = H$. Then $F([A, B]) = [F(A), F(B)] = K = [B, A]$ (Lemma C3). It follows from Lemma C2 that $F(A) = B$ and $F(B) = A$. We now have that Λ is the disjoint union $\Lambda = A \cup (\Lambda - H) \cup B \cup (\Lambda - K)$. If C is any of A, $\Lambda - H$, B, or $\Lambda - K$ then $F(C) \cap C = \varnothing$ and $F^2(C) = C$. It is now clear that every periodic point of F in Λ has even period and we have the situation of (ii). $\square$

PROPOSITION D2. *Suppose that G is an orientation-preserving homeomorphism of the plane with invariant cofrontier Λ. If G has a fixed point in Λ and P is a subcontinuum of Λ such that $P \cup G(P) = \Lambda$, then $P \cup G^k(P) = \Lambda$ for all $k \geq 1$.*

PROOF. Reference to Lemmas C2 and C3 is implicit in manipulations of the [,] notation in the argument that follows. If Λ is a circle the result is clear. The idea in the following proof is to mimic the circle situation with an arbitrary cofrontier Λ. The notation developed in §III(C) makes this possible.

If $P = \Lambda$ we have nothing to prove. If P is proper then, since $P \cup G(P) = \Lambda$, $\text{int}_\Lambda(P) \neq \varnothing$. We have, in this case, by Lemma A4, that $\text{cl}(\Lambda - P) \cap \text{cl}(\Lambda - \text{cl}(\Lambda - P)) = A \cup B$ with $\{A, B\}$ an irreducible separating pair in Λ. With the right choice of what we call A and what we call B (or with the right choice of generator of $\check{H}^1(\Lambda)$) we have $[A, B] = \text{cl}(\Lambda - \text{cl}(\Lambda - P))$ and $[B, A] = \text{cl}(\Lambda - P)$.

Now $\text{int}_\Lambda P \subset \Lambda - \text{cl}(\Lambda - P) = \Lambda - [B, A] = (A, B)$ and, since $P \cup G(P) = \Lambda$, $(\text{int}_\Lambda P) \cup G(\text{int}_\Lambda P)$ is dense in Λ. It follows that $(A, B) \cup G((A, B))$ is dense in Λ and that $\text{cl}((A, B) \cup G((A, B))) = [A, B] \cup G([A, B]) = \Lambda$. From this we see that the fixed points of G in Λ are in $[A, B]$. Also, since $[A, B] \cup G([A, B]) = \Lambda$ we have $G([A, B]) \supset \Lambda - [A, B] = (B, A)$ so that $G([A, B]) \supset \text{cl}((B, A)) = [B, A]$.

CLAIM 1. $G([A, B]) \cap [A, B] = [A, G(B)] \cup [G(A), B]$.

For the proof of this claim we need only note that $\{G(A), G(B)\}$ is an irreducible separating pair and that $[G(A), G(B)] \cup [A, B] = G([A, B]) \cup [A, B] = \Lambda$. Then, by Lemma C3,

$$G([A, B]) \cap [A, B] = [G(A), G(B)] \cap [A, B]$$
$$= [A, G(B)] \cup [G(A), B].$$

Now let p be a fixed point of G in P. Then $p \in [A, B]$ (since $G([A, B]) \cup [A, B] = \Lambda$) and $p \in [A, B] \cap G([A, B]) = [A, B] \cap [G(A), G(B)]$. Thus $p \in [G(A), B]$ or $p \in [A, G(B)]$. We assume, for the rest of the proof, that $p \in [G(A), B]$—the case $p \in [A, G(B)]$ is similar.

CLAIM 2. For all $k \geq 0$:

(i) $G^{k+1}(A) \subset [G^k(A), B] \subset [A, B]$;

(ii) $p \in [G^{k+1}(A), B]$; and

(iii) $G([G^k(A), B]) \supset [B, A]$.

We prove Claim 2 by induction on k. For $k = 0$, (i) follows from Claim 1, (ii) is our assumption on the location of p, and (iii) was established just prior to the statement of Claim 1.

Assume now that (i), (ii), and (iii) are correct for some $k = n \geq 0$. Then $[G^n(A), B] = [G^n(A), G^{n+1}(A)] \cup [G^{n+1}(A), B]$ so that $G([G^n(A), B]) = [G^{n+1}(A), G^{n+2}(A)] \cup [G^{n+2}(A), G(B)]$ since $G([G^n(A), B]) \supset [B, A]$, $B \subset G([G^n(A), B]) = [G^{n+1}(A), G(B)]$ and $[G^{n+1}(A), G(B)] = [G^{n+1}(A), B] \cup [B, G(B)]$. Now if $G^{n+2}(A) \not\subset [G^{n+1}(A), B]$ then $G^{n+2}(A) \subset (B, G(B))$ so that $G^{n+2}(A) \subset [G^{n+1}(A), G(B)]$. But then $(B, G(B)) = (B, G^{n+2}(A)) \cup [G^{n+2}(A), G(B)]$ and, in particular, $[G^{n+2}(A), G(B)] = G([G^{n+1}(A), B]) \subset (B, G(B))$. Since $p \notin (B, G(B)) \cap [G(A), B] = \varnothing$, $p \notin (B, G(B))$. On the other hand, $p = G(p) \in G([G^{n+1}(A), B]) \subset (B, G(B))$ so that p is in

$(B, G(B))$. Thus $G^{n+2}(A) \subset [G^{n+1}(A), B]$. Also, $[A, B] \supset [G^n(A), B] = [G^n(A), G^{n+1}(A)] \cup [G^{n+1}(A), B]$ so that $[A, B] \supset [G^{n+1}(A), B]$ and (i) is correct for $k = n + 1$.

We have

$$[B, A] \supset G([G^n(A), B]) = G([G^n(A), G^{n+1}(A)] \cup [G^{n+1}(A), B])$$
$$= [G^{n+1}(A), G^{n+2}(A)] \cup G([G^{n+1}(A), B])$$

and since $G^{n+2}(A) \subset [G^{n+1}(A), B]$ we also have $[G^{n+1}(A), G^{n+2}(A)] \subset [G^{n+1}(A), B] \subset [A, B]$. Thus $G([G^{n+1}(A), B]) \supset (B, A]$ so $G([G^{n+1}(A), B]) \supset \mathrm{cl}((B, A]) = [B, A]$ and (iii) is correct for $k = n + 1$.

Now $p \in [G^{n+1}(A), B]$ implies that $p = G(p) \in G([G^{n+1}(A), B]) = [G^{n+2}(A), G(B)] = [G^{n+2}(A), B] \cup (B, G(B)]$. That $p \in [G(A), B]$ and $[G(A), B] \cap (B, G(B)] = \varnothing$ now implies that $p \in [G^{n+2}(A), B]$ and (ii) is also correct for $k = n + 1$ and Claim 2 is proved.

To complete the proof of the proposition, suppose that $G^n([A, B]) \cup [A, B] = \Lambda$ for some $n \geq 1$. Then $[G^n(A), G^n(B)] \supset [B, A]$ so that $[G^n(A), G^n(B)] = [G^n(A), B] \cup [B, G(B)]$. Then $G^{n+1}([A, B]) \supset G([G^n(A), B]) \supset [B, A]$, by Claim 2, so that $G^{n+1}([A, B]) \cup [A, B] = \Lambda$. Thus, by induction, $G^k([A, B]) \cup [A, B] = \Lambda$ for all $k \geq 1$. Recall that $[A, B] = \mathrm{cl}(\Lambda - \mathrm{cl}(\Lambda - P)) \subset P$. Thus $G^k(P) \cup P = \Lambda$ for all $k \geq 1$. □

If $f: X \to X$ is a homeomorphism of the metric space X, a point $x \in X$ is *recurrent* under f provided there is a sequence of integers $n_k \to \infty$ such that $\lim_{k\to\infty} f^{n_k}(x) = x$. A theorem of Gottschalk (Theorem 1 of [G]) states that if x is recurrent under f then x is recurrent under f^n for all n.

PROOF OF THEOREM A. Let F be an orientation-preserving homeomorphism of the plane with invariant cofrontier Λ, suppose that $p \in \Lambda$ is fixed by F, and suppose that ρ_i, the rotation number of $\hat{F}_i: \mathscr{S}_i \to \mathscr{S}_i$, is nonzero (the case $\rho_e \neq 0$ is similar). Let $\mathscr{P}$ be a prime end of $\mathscr{D}_i$ with p in the impression, $I(\mathscr{P})$, of $\mathscr{P}$ (such a $\mathscr{P}$ exists since $\Lambda = \mathscr{F} \mathscr{D}_i$).

Now let m_k be a sequence of positive integers, $m_k \to \infty$, such that $\hat{F}_i^{m_k}(\mathscr{P})$ converges to a prime end $\mathbb{Q}$ as $k \to \infty$ ($\mathscr{S}_i$ is compact). Then $\mathbb{Q}$ is recurrent under $\hat{F}_i$ ($\mathbb{Q}$ is in the w-limit set of $\mathscr{P}$, points of the w-limit set are nonwandering, and nonwandering points under a circle homeomorphism are recurrent; see [vK]). Moreover, since $\lim_{k\to\infty} \hat{F}_i^{m_k}(\mathscr{P}) = \mathbb{Q}$, $\limsup F^{m_k}(I(\mathscr{P})) \subset I(\mathbb{Q})$ so that $p \in I(\mathbb{Q})$.

It follows now from Lemma B3 that, for some $M \geq 1$, $F^M(I(\mathbb{Q})) \cup (I(\mathbb{Q})) = \Lambda$. The homeomorphism $G = F^M$ is also orientation-preserving with invariant cofrontier Λ. Since $G(I(\mathbb{Q})) \cup (I(\mathbb{Q})) = \Lambda$ and G has a fixed point in $I(\mathbb{Q})$, Proposition D2 allows us to conclude that $G^k(I(\mathbb{Q})) \cup (I(\mathbb{Q})) = \Lambda$ for all $k \geq 1$. Since $\mathbb{Q}$ is recurrent under $\hat{F}_i$, $\mathbb{Q}$ is also recurrent under $\hat{F}_i^M = \hat{G}$. Let $n_k \to \infty$ be such that $\lim_{k\to\infty} \hat{G}^{n_k}(\mathbb{Q}) = \mathbb{Q}$. Then $\limsup G^{n_k}(I(\mathbb{Q})) \subset I(\mathbb{Q})$. But we also have that $G^{n_k}(I(\mathbb{Q})) \cup (I(\mathbb{Q})) = \Lambda$ for

$n_k \geq 1$. Thus

$$I(\mathbb{Q}) = (\limsup G^{n_k}(I(\mathbb{Q}))) \cup (I(\mathbb{Q})) = \limsup(G^{n_k}(I(\mathbb{Q})) \cup (I(\mathbb{Q}))) = \Lambda.$$

We now have a prime end, $\mathbb{Q}$, of $\mathscr{D}_i$, the impression of which is all of Λ. It follows from a theorem of Rutt [**R**, Theorem 2] that either Λ is indecomposable or $\Lambda = H \cup K$, H and K proper indecomposable subcontinua of Λ. We conclude from Proposition D1 that the latter cannot be the case since F has a fixed point in Λ and a nonzero prime end rotation number. $\square$

PROOF OF COROLLARY 1. Suppose that F is an orientation-preserving homeomorphism of the plane with invariant cofrontier Λ. Suppose that the rotation numbers ρ_i of $\hat{F}_i$ and ρ_e of $\hat{F}_e$ satisfy: $\rho_i = m/n$ is rational and ρ_e is irrational. Now $G = F^n$ is also an orientation-preserving homeomorphism of the plane with invariant cofrontier Λ. Since $\hat{G}_i = \hat{F}_i^n$ and $\hat{G}_e = \hat{F}_e^n$, the prime end rotation numbers of G are $\rho(\hat{G}_i) = n(m/n)$ (mod 1) $= 0$ and $\rho(\hat{G}_e) = n\rho_e$ (mod 1) $\neq 0$.

Since $\rho(\hat{G}_i) = 0$ there is a fixed prime end (under $\hat{G}_i$), say $\mathscr{P}$, of $\mathscr{D}_i$. Thus $G(I(\mathscr{P})) = I(\mathscr{P})$. If $I(\mathscr{P})$ is proper in Λ then, since $I(\mathscr{P})$ does not separate the plane, $I(\mathscr{P})$ contains a fixed point of G by the fixed point theorem of Cartwright and Littlewood (Theorem 5 of [**CL2**]). In this case G has a fixed point in Λ and a nonzero prime end rotation number $(n\rho_e$ (mod 1)). Thus, by Theorem A, Λ is indecomposable.

If $I(\mathscr{P}) = \Lambda$, then Λ is either indecomposable or is the union of two proper indecomposable subcontinua [**R**, Theorem 2]. In the latter case, since the prime end rotation numbers of G are not both $1/2$, G must have a fixed point in Λ by Proposition D1. Then again we may apply Theorem A to G to obtain that Λ is indecomposable.

To prove (ii) of the corollary, suppose that $\rho_i = m/n$ and $\rho_e = k/l$, both reduced to lowest terms, with $n \neq l$ (if m or k is zero we assume n, respectively l, is 1). Assume that n is the lesser of n and l and let $G = F^n$. Then, as in the proof above of (i), G induces prime end rotation numbers $\rho(\hat{G}_i) = n(m/n)$ (mod 1) $= 0$ and $\rho(\hat{G}_e) = n(k/l)$ (mod 1). If $n(k/l)$ (mod 1) $= 0$ then l divides nk. But l and k are relatively prime and $1 \leq n < l$ so this is impossible. Thus $\rho(\hat{G}_e) \neq 0$. It follows, as in the argument above for part (i), that Λ is indecomposable. $\square$

PROOF OF THEOREM B. Let F be an orientation-preserving homeomorphism of the plane with invariant cofrontier Λ. Suppose that F has no fixed points in Λ and that p and q are periodic points of F in Λ of periods m and n, respectively, with m and n relatively prime.

If F induces a zero prime end rotation number, say $\rho_i = 0$, then there is a prime end $\mathscr{P}$ of $\mathscr{D}_i$ with $\hat{F}_i(\mathscr{P}) = \mathscr{P}$. Then $F(I(\mathscr{P})) = I(\mathscr{P})$. If $I(\mathscr{P})$ is proper, $I(\mathscr{P})$ does not separate the plane and must contain a fixed point [**CL2**, Theorem 5]. But F has no fixed points in Λ. Thus $I(\mathscr{P}) = \Lambda$. Then Λ is indecomposable or $\Lambda = H \cup K$, H and K proper indecomposable subcontinua [**R**, Theorem 2]. In the latter case, since F has no fixed points in Λ, all periodic points of F in Λ must have even period (Proposition D1). This is not the case since m and n are relatively prime.

Now suppose that $\rho_i \neq 0$. It follows from the relative primeness of m and n that one of $m\rho_i$ (mod 1) and $n\rho_i$ (mod 1) is nonzero. Say $m\rho_i$ (mod 1) $\neq 0$. Letting $G = F^m$, G has a fixed pont in Λ (namely p) and $\rho(\hat{G}_i) = \rho(\hat{F}_i^m) = m\rho_i \neq 0$. If $n\rho_i$ (mod 1) $\neq 0$ let $G = F^n$. In either case Theorem A applies to G to yield the indecomposability of Λ. $\square$

PROOF OF COROLLARY 2. Assuming the hypotheses of the corollary, let l be the greatest common divisor of m and n and let $G = F^l$. Then G has no fixed points in Λ, and G has periodic points of periods m/l and n/l in Λ. Since m/l and n/l are relatively prime, Λ is indecomposable by Theorem B. $\square$

PROOF OF COROLLARY 3. Let T be the homeomorphism referred to in the discussion preceding the statement of Corollary 3. For the appropriate parameter values in the forced van der Pol equation, the following properties of T are established in [CL1] (see also [L]):

   (i) T is an orientation-preserving homeomorphism of the plane;
  (ii) the origin, 0, is a stable fixed point of T^{-1} and T has no other (finite) fixed points;
 (iii) the point at infinity, ∞, is a stable fixed point of T^{-1}.

Let U and V be the basins of 0 and ∞, respectively, under T^{-1}. That is,

$$U = \{x \in \mathbb{R}^2 | T^{-n}(x) \to 0 \text{ as } n \to \infty\}$$

and

$$V = \{x \in \mathbb{R}^2 | T^{-n}(x) \to 0 \text{ as } n \to \infty\}.$$

Then (also from [CL1]):

  (iv) there are stable, under T, periodic points of periods $2n - 1$ and $2n + 1$ (some $n \geq 2$) and these points are in $\text{cl}(U) \cap \text{cl}(V)$;
  (v) $U \cup V$ is dense in $\mathbb{R}^2$.

Now let $F^* = \text{cl}(U) \cap \text{cl}(V)$. F^* is then an invariant plane-separating continuum containing (by (iv)) periodic points of periods $2n - 1$ and $2n + 1$ and containing no fixed points. If W is a complementary domain of F^* then $W \cap U \neq \varnothing$ or $W \cap V \neq \varnothing$ by (v) so that W is either the component of $\mathbb{R}^2 - F^*$ containing U or the component of $\mathbb{R}^2 - F^*$ containing V. That is, F^* has precisely two complementary domains, $\mathscr{D}_i \supset U$ and $\mathscr{D}_e \supset V$. Since $F^* = \text{cl}(U) \cap \text{cl}(V)$, $F^* = \mathscr{F} \mathscr{D}_i = \mathscr{F} \mathscr{D}_e$ and it follows that F^* is a cofrontier (see Lemma 41 of [CL2]). Theorem B now applies ($2n - 1$ and $2n + 1$ are relatively prime) to prove that F^* is an indecomposable continuum. $\square$

References

[AS] K. T. Alligood and T. Saurer, *Rotation numbers of periodic orbits in the Hénon map*, Preprint.

[B] M. Barge, *Homoclinic intersections and indecomposability*, Proc. Amer. Math. Soc. **101** (1987), 541–544.

[BG] M. Barge and R. Gillette, *Rotation in invariant plane separating continua*, Preprint.

[Bi] G. D. Birkhoff, *Sur quelques courbes fermees remarquables*, Bull. Soc. Math. France **60** (1932), 1–26.

[C] C. Caratheodory, *Uber die Begrenzung einfach zusammenhangender Gebiete*, Math. Ann. **73** (1913), 323–370.

[Ch] M. Charpentier, *Sur quelques propriétés des courbes de M. Birkhoff*, Bull. Soc. Math. France **62** (1934), 193–224.

[CL1] M. L. Cartwright and J. E. Littlewood, *On non-linear differential equations of the second order.* I, *The equation* $\ddot{y} - k(1 - y^2)\dot{y} + y = b\lambda k \cos(\lambda t + \alpha)$, *k large*, J. London Math. Soc. **20** (1945), 180–189.

[CL2] ____, *Some fixed point theorems*, Ann. Math. **54** (1951), 1–37.

[D] A. Denjoy, *Sur les courbes définies par les équations differentielles à la surface due tore*, J. Math. Pures Appl. **11** (1932), 42–49.

[G] W. H. Gottschalk, *Powers of homeomorphisms with almost periodic properties*, Bull. Amer. Math. Soc. **50** (1944), 222–227.

[Gr] M. J. Greenberg, *Lectures on Algebraic Topology*, Math. Lecture Note Ser., W. A. Benjamin, Reading, Mass., 1967.

[HH] K. Hockett and P. Holmes, *Josephson's junction, annulus maps, Birkhoff attractors, horseshoes and rotation sets*, Ergodic Theory Dynamical Systems **6** (1986), 205–239.

[HW] W. Hurewicz and H. Wallman, *Dimension theory*, Princeton Univ. Press, Princeton, N.J., 1948.

[HY] J. Hocking and G. Young, *Topology*, Addison-Wesley, Reading, Mass., 1961.

[K1] K. Kuratowski, *Topology*, vol. II, Academic Press, New York and London, 1968.

[K2] ____, *Sur les coupres irréductibles du plan*, Fund. Math. **6** (1924), 130–145.

[Kn] B. Knaster, *Quelques coupres singulières du plan*, Fund. Math. **7** (1925), 264–289.

[L] M. Levi, *Qualitative analysis of the periodically forced relaxation oscillations*, Memoirs, vol. 214, Amer. Math. Soc., Providence, R.I., 1981, pp. 1–147.

[LeC] P. Le Calvez, *Propriétés des attracteurs de Birkhoff*, Ergodic Theory Dynamical Systems **8** (1988), 241–310.

[Ln] N. Levinson, *A second order differential equation with singular solutions*, Ann. Math. **50** (1949), 127–153.

[M] J. Mather, *Topological proofs of some purely topological consequences of Caratheodory's theory of prime ends* (Th. M. Rassias, G. M. Rassias, eds.), Slected Studies, North-Holland, Amsterdam, 1982, pp. 225–255.

[N] M. H. A. Newman, *Topology of plane sets*, 2nd ed., Cambridge Univ. Press, 1954.

[Ni] Z. Nitecki, *Differentiable dynamics*, M.I.T. Press, Cambridge, Mass. and London, 1971.

[P] R. Plykin, *Sources and sinks for A-diffeomorphisms*, USSR Math. Sb. **23** (1974), 233–253.

[R] N. E. Rutt, *Prime ends and indecomposability*, Bull. Amer. Math. Soc. **41** (1935), 265–273.

[S] E. Spanier, *Algebraic topology*, McGraw-Hill, New York, 1966.

[vK] E. R. van Kampen, *The topological transformations of a simple closed curve into itself*, Amer. J. Math. **57** (1935), 142–152.

[W] R. Walker, *Basin boundaries with irrational rotation on prime ends*, Preprint.

Department of Mathematical Sciences, Montana State University, Bozeman, Montana 59717

Contemporary Mathematics
Volume **117**, 1991

Rotational Dynamics on Cofrontiers

B. L. BRECHNER, M. D. GUAY, AND J. C. MAYER

ABSTRACT. A *pseudorotation* is defined to be an extendable homeomorphism of a cofrontier $\Lambda \subset \mathbf{R}^2$ whose induced prime end homeomorphisms from both sides are conjugate to a single rotation R_α on S^1. A class $\mathcal{H}$ of cofrontiers in the plane is constructed following the recipe suggested by M. Handel [**Ha**]. The authors study periodicity, almost periodicity, and recurrence properties of pseudorotations of cofrontiers $\Lambda \in \mathcal{H}$, as well as cofrontiers in general. They also study the relationship between each of these properties on Λ and the corresponding property on the induced prime end circles and disks.

Introduction

We construct a class $\mathcal{H}$ of cofrontiers in the plane, together with a class of autohomeomorphisms which we call pseudorotations. These examples are suggested by Handel's construction in [**Ha**]. We are interested in studying both the topological and the embedding-dependent topological properties of these cofrontiers as described by their prime end structure, as well as the dynamical properties of the pseudorotations.

Initially, some (corollaries of) theorems of Hemmingsen [**He**] and Cartwright [**Ca**] are observed, characterizing (1) invariant cofrontiers under irrational rotations of the plane and (2) a minimal cofrontier under an almost periodic homeomorphism of the cofrontier, respectively (Theorems 1 and 3), and relating almost periodicity on an arbitrary cofrontier-plus-interior with almost periodicity on its prime end circle (Theorem 2). Theorems 4 and 5 are prime end theorems, about periodic pseudorotations and the prime end

1980 *Mathematics Subject Classification* (1985 *Revision*). Primary 54F20, 54F50, 54G15, 54G20, 54H20.

Key words and phrases. Almost periodicity, cofrontiers, dynamics, equicontinuity, indecomposable continua, irrational numbers of constant type, minimal set, plane homeomorphisms, prime ends, recurrence, rotations.

Supported in part by NSF-Alabama EPSCoR grant number RII-8610669.

The final version of this paper will be submitted for publication elsewhere.

structure of indecomposable cofrontiers admitting irrational pseudorotations, respectively.

Next, we construct the family $\mathscr{H}$ carefully, and obtain some topological, dynamical, and number-theoretic results about cofrontiers $\Lambda \in \mathscr{H}$ and pseudorotations f_α of Λ. (f_α is a pseudorotation on Λ if and only if the induced rotation on the circle of prime ends on each side of Λ is of angle $2\pi\alpha$ for some $\alpha \in [0, 1)$.) In Theorem 9, we prove that *every* $\Lambda \in \mathscr{H}$ admits an irrational pseudorotation f_α, for some α. In Theorem 14, we put restrictions on several of the technical conditions in the construction of Λ to show that such restricted Λ are indecomposable, and if f_α is a "standard" pseudorotation on such a restricted $\Lambda \in \mathscr{H}$, then α is irrational and not of constant type, and f_α is recurrent, minimal, and not almost periodic on Λ.

The main theorems of this paper are Theorems 8 and 14. Theorem 8 describes the prime end structure of the indecomposable members of our family $\mathscr{H}$ (this follows from Theorems 5 and 9), while Theorem 14 describes the dynamical properties and a number-theoretic property of certain "restricted" indecomposable members of $\mathscr{H}$. Theorem 14 follows from Theorem 6 and Theorems 9 through 13.

Cofrontiers, prime ends, and pseudorotations

Cofrontier. A *cofrontier* Λ is a continuum in the plane $\mathbf{R}^2$ which irreducibly separates $\mathbf{R}^2$ into exactly two components. That Λ separates $\mathbf{R}^2$ into exactly two components is a topological property, as is the property that a cofrontier is the common boundary of its complementary domains. Two cofrontiers Λ_1 and Λ_2 are said to be *equivalent* iff there is a homeomorphism $h : (\mathbf{R}^2, \Lambda_1) \to (\mathbf{R}^2, \Lambda_2)$. A cofrontier may well have inequivalent embeddings in $\mathbf{R}^2$ (though S^1 does not, of course). We usually denote a cofrontier together with its embedding into $\mathbf{R}^2$, by Λ. We denote the unbounded component of $\mathbf{R}^2 \setminus \Lambda$ by $\mathrm{Ext}(\Lambda)$ and the bounded component by $\mathrm{Int}(\Lambda)$.

Prime end compactification. Suppose Λ is a cofrontier and $f : (\mathbf{R}^2, \Lambda) \to (\mathbf{R}^2, \Lambda)$ is a homeomorphism. The *prime end compactification* $\phi_i : \mathrm{Int}(\Lambda) \to D_i$ (where D_i is a copy of the open unit disk), and the homeomorphism $f|\mathrm{Int}(\Lambda)$ induce a homeomorphism $F_i : \mathrm{Cl}(D_i) \to \mathrm{Cl}(D_i)$ as the extension of $\phi_i f \phi_i^{-1}$. We call $F_i|\mathrm{Bd}(D_i)$ the induced homeomorphism on the circle $\mathrm{Bd}(D_i)$ of prime ends. ϕ_i may be taken to be a conformal homeomorphism, and it exists by the Riemann Mapping Theorem. (See Figure 1.) Similarly, the prime end compactification $\phi_e : \mathrm{Ext}(\Lambda) \cup \{\infty\} \to D_e$ (where D_e is a copy of the open unit disk), and the homeomorphism $f|\mathrm{Ext}(\Lambda)$ induce a homeomorphism $F_e : \mathrm{Cl}(D_e) \to \mathrm{Cl}(D_e)$ as the extension of $\phi_e f \phi_e^{-1}$. We call $F_e|\mathrm{Bd}(D_e)$ the induced homeomorphism on the circle $\mathrm{Bd}(D_e)$ of prime ends. In general, neither $F_i|\mathrm{Bd}(D_i)$ nor $F_e|\mathrm{Bd}(D_e)$ need be conjugate to any rotation of S^1, let alone the same rotation.

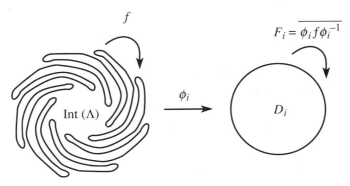

FIGURE 1

Pseudorotation. A *pseudorotation* of a cofrontier Λ is a homeomorphism $f : \Lambda \to \Lambda$, extendable to a homeomorphism of $\mathbf{R}^2$, such that each of the induced homeomorphisms on the circles of prime ends, corresponding respectively to $\mathrm{Int}(\Lambda)$ and $\mathrm{Ext}(\Lambda)$, is conjugate to the same rotation R_α of S^1 through angle $2\pi\alpha$ for some $\alpha \in [0, 1)$. Since this α is uniquely determined, we write f_α when f is a pseudorotation and R_α is the induced rotation on the circle of prime ends. Since f_α is extendable, we often assume a particular extension $f_\alpha : (\mathbf{R}^2, \Lambda) \to (\mathbf{R}^2, \Lambda)$. A given pseudorotation will have many extensions to homeomorphisms.

Recurrent, almost periodic, and minimal homeomorphisms

A homeomorphism $h : X \to X$ is said to be *recurrent* iff for every $\epsilon > 0$, there is an $n \in \mathbf{Z}$ such that $\mathrm{d}(h^n, \mathrm{id}_X) < \epsilon$. Equivalently, h is recurrent iff $\{h^n | n \in \mathbf{Z}\}$ has a subsequence converging uniformly to the identity.

A homeomorphism $h : X \to X$ is said to be *almost periodic* iff for every $\epsilon > 0$ there is a subsequence $\{h^{n_i} | i \in \mathbf{Z}\}$ of $\{h^n | n \in \mathbf{Z}\}$ with bounded gaps such that for all i, $\mathrm{d}(h^{n_i}, \mathrm{id}_X) < \epsilon$. On a compact metric space, h is almost periodic iff $\{h^n | n \in \mathbf{Z}\}$ is an equicontinuous family of iterates. Clearly, h periodic implies h almost periodic implies h recurrent. Recall that the recurrent homeomorphisms of S^1 are almost periodic, and the almost periodic homeomorphisms of S^1 are conjugate to rotations, with the rational rotations being periodic and the irrational rotations being nonperiodic, almost periodic. In particular, recurrent is equivalent to almost periodic on S^1.

A homeomorphism $h : X \to X$ is *minimal* iff X is the only nonempty closed invariant subset of X under h. It follows that $h : X \to X$ is minimal iff every orbit is dense. Recall that an irrational rotation of S^1 is minimal.

In [**He**], Hemmingsen shows that topological disks and annuli are the only plane continua with interior points which (continua) admit nonperiodic, almost periodic homeomorphisms (i.e., homeomorphisms whose families of iterates are equicontinuous), and that any such homeomorphism must be conjugate to an irrational rotation on the disk or annulus. The first two theorems below may be seen as corollaries of Hemmingsen's theorem, although the first one is very easy to prove directly. The third is an observation of Cartwright [**Ca**] based on Hemmingsen's results.

THEOREM 1. *Let* $h : \mathbf{R}^2 \to \mathbf{R}^2$ *be conjugate to an irrational rotation, and let* Λ *be a cofrontier. If* $h(\Lambda) = \Lambda$*, then* Λ *is homeomorphic to* S^1*.*

THEOREM 2. *Let* h *be an almost periodic, nonperiodic homeomorphism of* $\Lambda \cup \mathrm{Int}(\Lambda)$ *onto itself, where* Λ *is a cofrontier in* $\mathbf{R}^2$*. Then* Λ *is a simple closed curve and* $h|\Lambda$ *is conjugate to an irrational rotation. Thus, the induced homeomorphism on the prime end disk will also be conjugate to an irrational rotation, and therefore almost periodic on the circle of prime ends.*

THEOREM 3. *Let* $h : (\mathbf{R}^2, \Lambda) \to (\mathbf{R}^2, \Lambda)$ *be a homeomorphism, where* Λ *is a nondegenerate continuum. If* h *is both almost periodic and minimal on* Λ*, then* Λ *is a simple closed curve and* $h|\Lambda$ *is conjugate to an irrational rotation.*

We obtain the following additional theorems about cofrontiers and pseudorotations in general. The forward direction of Theorem 4 is a corollary of Theorem 3.14 (and its proof) in [**Br**].

THEOREM 4 (PERIODIC PSEUDOROTATIONS). *Suppose* Λ *is a cofrontier and* $h : (\mathbf{R}^2, \Lambda) \to (\mathbf{R}^2, \Lambda)$ *is a homeomorphism. Let* $H : \mathrm{Cl}(D) \to \mathrm{Cl}(D)$ *denote the induced homeomorphism on either the interior or exterior prime end compactification. Then* $h|\Lambda$ *is periodic iff* $H|\mathrm{Bd}(D)$ *is periodic.*

The following theorem, which follows from a theorem of Rutt [**Ru**], shows that the fine structure of an indecomposable cofrontier admitting an irrational pseudorotation must be quite complicated. Rogers [**Ro**] has obtained a theorem similar to Theorem 5 for extendable intrinsic rotations of a bounded domain.

THEOREM 5. *Suppose* Λ *is a cofrontier which admits an irrational pseudorotation. Then the following are equivalent:*

(1) Λ *is indecomposable.*
(2) *Every prime end of* $\mathrm{Int}(\Lambda)$ *(respectively,* $\mathrm{Ext}(\Lambda)$*) has as its impression all of* Λ*.*
(3) *Every pair of points of* Λ *accessible from* $\mathrm{Int}(\Lambda)$ *(respectively,* $\mathrm{Ext}(\Lambda)$*) are in different composants.*
(4) *The set of prime ends of* $\mathrm{Int}(\Lambda)$ *(respectively,* $\mathrm{Ext}(\Lambda)$*) corresponding to simple dense canals is second category (thus, dense) in the circle of prime ends.*

Thus, we are interested in looking at a class of cofrontiers, $\mathscr{H}$, including indecomposable cofrontiers, which are constructed in a regular fashion, and which are a potential source of examples and counterexamples.

Construction of the family $\mathscr{H}$ of cofrontiers

A cofrontier $\Lambda \in \mathscr{H}$ is constructed as the intersection $\Lambda = \bigcap_{k=1}^{\infty} A_k$ of annuli A_k, where A_{k+1} is embedded essentially in A_k and is invariant under certain rotations of A_k. Refer to Figure 2 for an example of the construction outlined below.

(1) A_k is divided into q_k links of diameter ϵ_k.

(2) The embedding of A_{k+1} in A_k is induced by lifting an embedding of an annulus $\bar{A}_{k+1}$ in a "single link" annulus A (the image of A_k under a q_k–to–one covering map).

(3) A *basic repetitive unit* of A_{k+1} *spans* some number, say s_{k+1}, of links of A_k. (In the example, $s_{k+1} = 2$.)

(4) The "fractional part" of A_k that a basic repetitive unit of A_{k+1} spans is $c_{k+1} = s_{k+1}/q_k$. (In the example, $c_{k+1} = 1/4$.)

(5) Each segment of A_{k+1} that spans a link of A_k is divided into $r_{k+1} > 1$ links. This can be done via the lift. (In the example, $r_{k+1} = 2$.)

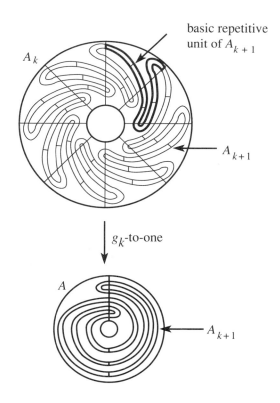

FIGURE 2

Inducing pseudorotations on $\Lambda \in \mathscr{H}$

We induce a pseudorotation f_α on $\Lambda \in \mathscr{H}$ as the uniform limit of topological rotations f_k of the annuli A_k. Refer to Figure 3 for an example.

(1) Homeomorphism f_1 rotates A_1 onto itself by the fractional amount m_1/q_1, taking links to links, and leaving A_2 invariant.

(2) Homeomorphism f_2 first does f_1, then rotates A_2 onto itself by the amount m_2/q_2 within its own annular structure, taking links to links, and leaving A_3 invariant. Thus, f_2 rotates A_2 the total amount $(m_1/q_1) + (m_2/q_2)$.

(3) In general, homeomorphism f_k first does f_{k-1}, then rotates A_k an additional amount m_k/q_k within its own annular structure. Thus, f_k rotates A_k the total amount

$$\frac{p_k}{q_k} = \sum_{j=1}^{k} \frac{m_j}{q_j}.$$

(4) Assume that the m_k's are not too large, i.e., $m_k < r_k$, and that the ϵ_k's are summable.

(5) In the limit,

$$\frac{p_k}{q_k} \to \alpha = \sum_{j=1}^{\infty} \frac{m_j}{q_j}$$

and the homeomorphisms f_k converge uniformly to the pseudorotation f_α.

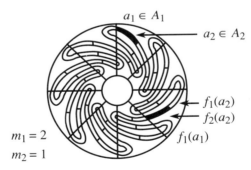

$a_1 \in A_1$

$a_2 \in A_2$

$f_1(a_2)$

$f_2(a_2)$

$f_1(a_1)$

$m_1 = 2$

$m_2 = 1$

FIGURE 3

We call any pseudorotation on Λ, induced as above, a *standard* pseudorotation.

Properties of cofrontiers $\Lambda \in \mathscr{H}$

The "length" of basic repetitive units determines whether Λ is indecomposable or homeomorphic to S^1. There are two measures of this length: s_{k+1}, the absolute number of links of A_k that a basic repetitive unit of A_{k+1} spans, and c_{k+1}, the relative amount of A_k that a basic repetitive unit of A_{k+1} spans.

THEOREM 6 (Λ INDECOMPOSABLE). *If* $\sum_{k=2}^{\infty} c_k = \infty$, *then* Λ *is indecomposable.*

THEOREM 7 ($\Lambda \cong S^1$). *If the* s_k's *are uniformly bounded above, then* Λ *is homeomorphic to* S^1.

We note that in Theorem 9 below, we prove that every $\Lambda \in \mathscr{H}$ admits an irrational pseudorotation. Thus it follows that *every indecomposable member of* $\mathscr{H}$ *satisfies the conclusions of Theorem 5.* We restate this theorem for completeness:

THEOREM 8 (FIRST MAIN THEOREM). *Suppose* $\Lambda \in \mathscr{H}$. *Then the following are equivalent*:

(1) Λ *is indecomposable.*
(2) *Every prime end of* Int (Λ) *(respectively,* Ext (Λ)*) has as its impression all of* Λ.
(3) *Every pair of points of* Λ *accessible from* Int (Λ) *(respectively,* Ext (Λ)*) are in different composants.*
(4) *The set of prime ends of* Int (Λ) *(respectively,* Ext (Λ)*) corresponding to simple dense canals is second category (thus, dense) in the circle of prime ends.*

Properties of pseudorotations

Assume f_α is a standard pseudorotation induced on $\Lambda \in \mathscr{H}$. We obtain the following theorems detailing some of the dynamical properties of these standard pseudorotations.

THEOREM 9 (α IRRATIONAL). *Assume* f_α *is a standard pseudorotation on* $\Lambda \in \mathscr{H}$. *If* Λ *is not homeomorphic to* S^1 *(e.g.,* Λ *is indecomposable) and for infinitely many values of* k, $m_k > 0$, *then* α *is irrational. Thus every* $\Lambda \in \mathscr{H}$ *admits an irrational pseudorotation.*

THEOREM 10 (f_α RECURRENT). *Assume* f_α *is a standard pseudorotation on* $\Lambda \in \mathscr{H}$. *If for some* $M > 0$ *and for all* k, $r_{k+1} \geq q_k$ *and* $m_k \leq M$, *then* $f_\alpha|\Lambda$ *is recurrent.*

That is, if the number of subdivisions of a spanning segment is large, and the incremental rotations (m_k/q_k) are small, then f_α is recurrent *on* Λ.

Irrationals of constant type. The irrational α is of *constant type* iff there is an $\epsilon > 0$ such that for all rationals p/q in lowest terms,

$$|\alpha - \frac{p}{q}| \geq \frac{\epsilon}{q^2}.$$

Thus, in an informal sense, α is not well approximated by rationals.

THEOREM 11 (α NOT OF CONSTANT TYPE). *If*

(1) Λ *is not homeomorphic to* S^1,
(2) *for infinitely many values of* k, $m_k > 0$,
(3) *for some* $M > 0$ *and for all* k, $m_k \leq M$, *and*
(4) *for all* k, $r_{k+1} \geq q_k$,

then the irrational α *is not of constant type.*

The following example serves two purposes: it shows that the equivalence of almost periodic and recurrent does not hold for all cofrontiers, and it shows that Theorem 4 cannot be extended to almost periodic.

EXAMPLE (f_α NOT ALMOST PERIODIC). *Follow the construction recipe:*

(1) $q_1 = 4$,
(2) $c_{k+1} = \frac{1}{2}$ *for all* $k \geq 1$, *and*
(3) $r_{k+1} = q_k$ *for all* $k \geq 1$,

with minimal bending.

The cofrontier $\Lambda \in \mathcal{H}$ obtained will be indecomposable. Induce a pseudorotation f_α on Λ by choosing $m_k = 1$ for all k. Then f_α will be recurrent on Λ by Theorem 10, and α will be an irrational not of constant type. By Theorem 13 below, f_α is minimal on Λ. Then it follows that f_α is *not* almost periodic on Λ by Theorem 3. This phenomenon is quite general in $\mathcal{H}$, as the following theorem shows:

THEOREM 12 (f_α RECURRENT, NOT ALMOST PERIODIC). *Suppose* f_α *is a standard pseudorotation of* $\Lambda \in \mathcal{H}$. *If* $c_{k+1} \geq c > 0$, $r_{k+1} \geq q_k$, *and* $0 < m_k \leq M$ *for all* k, *then* f_α *is recurrent, but not almost periodic on* Λ.

THEOREM 13 (Λ MINIMAL UNDER f_α). *Suppose* f_α *is a standard pseudorotation of* $\Lambda \in \mathcal{H}$. *If* $r_{k+1} \geq q_k$ *and* $0 < m_k \leq M$ *for all* k, *then* f_α *is minimal on* Λ.

That is, if the number of subdivisions of a spanning segment is large, and the incremental rotations are small, then every orbit of $f_\alpha|\Lambda$ is dense in Λ.

THEOREM 14 (SECOND MAIN THEOREM). *Suppose* f_α *is a standard pseudorotation on* $\Lambda \in \mathcal{H}$. *Suppose further that*

(a) *there exists* $c > 0$ *such that for all* k, $c_{k+1} \geq c$,
(b) *for all* k, $r_{k+1} \geq q_k$,
(c) *for some* $M > 0$ *and for all* k, $0 < m_k \leq M$.

Then

(1) Λ *is indecomposable, by* (a) *and Theorem 6,*
(2) Λ *is a minimal set under* f_α, *by* (b), (c), *and Theorem 13,*
(3) f_α *is recurrent, by* (b), (c), *and Theorem 10,*

(4) f_α *is not almost periodic, by* (a), (b), (c), *and Theorem* 12, *or by conclusions* (1) *and* (2) *and Theorem* 3,

(5) α *is irrational, by* (a), (c), *and Theorem* 9, *and*

(6) α *is not of constant type, by* (a), (b), (c), *and Theorem* 11.

Questions

We conclude with some questions about cofrontiers and pseudorotations, both arbitrary ones and those in our class $\mathscr{H}$.

QUESTION 1. Does there exist an indecomposable cofrontier Λ admitting an irrational pseudorotation h_α for which α is of constant type?

QUESTION 2. Suppose Λ is a cofrontier and h_α is an irrational pseudorotation of Λ. Must $h_\alpha|\Lambda$ be minimal? Recurrent?

Note that M. Barge has asked the "minimal" part. An example due to Barge, the "split hairy circle," shows that the question should be asked for a pseudorotation, and not merely a one-sided pseudorotation.

QUESTION 3. Let $\mathscr{A}$ be the set of all irrationals α such that there exists an indecomposable cofrontier Λ (in $\mathscr{H}$ or arbitrary) and a pseudorotation h_α on Λ. Is $\mathscr{A}$ a proper subset of the irrationals? Does there exist a number-theoretic characterization of $\mathscr{A}$?

QUESTION 4. Are there any cofrontiers in $\mathscr{H}$ that are neither indecomposable, nor homeomorphic to S^1?

In an attempt to answer Question 4, one might try to construct a $\Lambda \in \mathscr{H}$ for which $\limsup s_k = \infty$ and $\sum_{k=2}^\infty c_k < \infty$ (so neither Theorem 6 nor 7 applies).

Recurrent and analytic extensions. We have not indicated explicitly how to extend the induced pseudorotations we obtain to $\mathbf{R}^2$. There are many possible extensions. One such procedure is outlined in Handel's construction of a diffeomorphism of $\mathbf{R}^2$ leaving a cofrontier Λ ($\cong$ pseudocircle) invariant [Ha]. Similar procedures can be applied in the authors' constructions. However, neither Handel's nor the authors' extensions are recurrent on $\text{Int}(\Lambda)$, nor conjugate on $\text{Int}(\Lambda)$ to the rotation R_α on the open unit disk D.

QUESTION 5. Suppose f_α is a standard pseudorotation on $\Lambda \in \mathscr{H}$, and that $f_\alpha|\Lambda$ is recurrent. Can $f_\alpha|\Lambda$ be extended so as to be recurrent on $\text{Int}(\Lambda)$? Can $f_\alpha|\Lambda$ be extended to $\text{Int}(\Lambda)$ so as to be conjugate to the rotation R_α on the open disk D?

M. R. Herman [Hr] obtains pseudorotations of some cofrontiers in S^2 that are complex analytically conjugate on $\text{Int}(\Lambda)$ to R_α on the complex unit disk $\mathbf{D}$ (and C^∞ conjugate on $\text{Ext}(\Lambda)$).

QUESTION 6. In general, when can a pseudorotation of a cofrontier $\Lambda \subset S^2$ be extended so as to be recurrent on $\text{Int}(\Lambda)$ (and $\text{Ext}(\Lambda)$)? (Complex analytically) conjugate on $\text{Int}(\Lambda)$ to R_α on $\mathbf{D}$?

REFERENCES

[Br]   B. Brechner, *Extendable periodic homeomorphisms on chainable continua*, Houston J. Math. **7** (1981), 327–344.

[Ca]   M. L. Cartwright, *Equicontinuous mappings of plane minimal sets*, Proc. London Math. Soc. (3) **14A** (1965), 51–54.

[Ha]   M. Handel, *A pathological area-preserving diffeomorphism of the plane*, Proc. Amer. Math. Soc. **86** (1982), 163–168.

[He]   E. Hemmingsen, *Plane continua admitting nonperiodic autohomeomorphisms with equicontinuous iterates*, Math. Scand. **2** (1954), 119–141.

[Hr]   M. R. Herman, *Construction of some curious diffeomorphisms of the Riemann sphere*, J. London Math. Soc. (2) **34** (1986), 375–384.

[Ro]   J. T. Rogers, Jr., *Intrinsic rotations of simply connected regions and their boundaries*, Preprint.

[Ru]   N. E. Rutt, *Prime ends and indecomposability*, Bull. Amer. Math. Soc. **41** (1935), 265–273.

DEPARTMENT OF MATHEMATICS, UNIVERSITY OF FLORIDA, GAINESVILLE, FL 32611

DEPARTMENT OF MATHEMATICS, UNIVERSITY OF SOUTHERN MAINE, PORTLAND, ME 04103

DEPARTMENT OF MATHEMATICS, UNIVERSITY OF ALABAMA AT BIRMINGHAM, BIRMINGHAM, AL 35294

Contemporary Mathematics
Volume **117**, 1991

Fundamental Regions of Planar Homeomorphisms

MORTON BROWN

1. Introduction

In his seminal paper [A] Steve Andrea revived and enriched the Brouwer theory of planar homeomorphisms. The principal new notion introduced in Andrea's paper was that of *fundamental region.*

The fundamental regions are useful invariants with which to distinguish various conjugacy invariant properties of Brouwer homeomorphisms. In this paper we shall briefly review Brouwer's theory and describe the Brouwer *translation domains* of some Brouwer homeomorphisms. We will then review Andrea's work and describe Brouwer homeomorphisms with a variety of different kinds of *fundamental regions.* We will introduce some techniques for constructing homeomorphisms with peculiar fundamental region structures. Finally we will present a very simple description of a Brouwer homeomorphism which is not invariant on any closed line.

2. The results of Brouwer

We restate the main results of Brouwer [**BRWR**] in a modern context. A fixed point free orientation-preserving homeomorphism of the plane will be called a Brouwer homeomorphism. A homeomorphism h of the plane (possibly with fixed points) is *free* [**B**] provided that whenever C is a compact connected set and $C \cap h(C) = \varnothing$ then for every nonzero integer n, $C \cap h^n(C) = \varnothing$.

BROUWER'S TRANSLATION LEMMA. *Let h be a Brouwer homeomorphism and let E be an arc in the plane with endpoints p and q where $q = h(p)$. Suppose $(E - q) \cap h(E - q) = \varnothing$. Then for all nonzero integers n, $(E - q) \cap h^n(E - q) = \varnothing$, and thus the orbit of $E - q$ is a topological line invariant under h.*

1980 *Mathematics Subject Classification* (1985 *Revision*). Primary 54F15, 58F99.

This paper is in final form and no version of it will be submitted for publication elsewhere.

Such a line is called a *translation line* (and E is called a *translation arc*). This line is not necessarily a closed subset of the plane (see Example 1, Figure 1c). If it *is* closed in the plane then it is called a *closed* translation line. As direct consequences of the Translation Lemma we have Propositions 1 and 2.

PROPOSITION 1. *Every Brouwer homeomorphism is free* [**BRWR, A, B**].

PROPOSITION 2. *If h is a Brouwer homeomorphism and p is a point in the plane, then the orbit of p converges (in both directions) to infinity.*

BROUWER'S TRANSLATION THEOREM [**BRWR**]. *Let h be a Brouwer homeomorphism and let p be a point of the plane. Let T be the translation $T(x, y) = (x + 1, y)$. Then there is a homeomorphism g of the plane such that $g(1/2, 0) = p$ and for all (x, y) such that $0 \le x \le 1$, $gt(x, y) = hg(x, y)$.*

In other words, each point lies in a topological infinite strip embedded as a closed set in the plane and which is "translated" by h to an adjacent strip. The union D of the orbits of the points of the strip (i.e., the orbit of the strip) is called a Brouwer *translation domain* of the homeomorphism h. Clearly, h acts on D like a translation. That is, there is a homeomorphism k of the plane onto D such that for all (x, y), $kt(x, y) = hk(x, y)$.

3. Remarks

Brouwer's *Translation Lemma* is a basic tool in the study of planar homeomorphisms. An immediate consequence of great importance is that an orientation-preserving homeomorphism of the plane with a periodic point necessarily has a fixed point. Brouwer's *Translation Theorem*, while much deeper and far more difficult to prove, has been used infrequently. One of the problems is that the *translation domain* may not fill the whole plane. A more serious difficulty is that the translation strip corresponding to a point p might necessarily transversely intersect a translation strip of a point q. The relationship between translation lines and translation strips is not what one might be led to expect from the definitions. In Brouwer's proof of the *Translation Theorem*, the *translation domain* is constructed as the orbit of a translation strip which is constructed transverse to an arbitrarily given *translation line*. The following example will help illustrate these comments.

4. Example 1. The "Reeb homeomorphism"

The homeomorphism h described in Figure 1a is the classical example of a Brouwer homeomorphism that is not topologically equivalent to a translation of the plane. Sometimes called (anachronistically) the two-dimensional Reeb homeomorphism, it was probably known to Brouwer, and it appears in Kerekjarto's book [**K**]. Figure 1a pictures a collection of oriented translation lines. The homeomorphism h moves points along the translation lines by

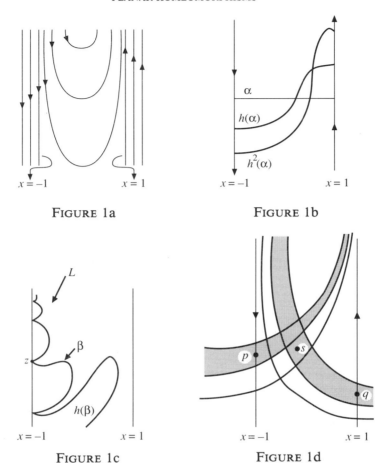

FIGURE 1a

FIGURE 1b

FIGURE 1c

FIGURE 1d

an arc length of 1. The homeomorphism h is not conjugate to a translation: Let α be the arc depicted in Figure 1b. For each n, $h(\alpha) \cap \alpha \neq \varnothing$, by the intermediate value theorem. If h were conjugate to a translation, the iterates of α would necessarily converge to infinity.

There are other translation lines than those pictured in Figure 1a. Figure 1c illustrates a translation arc, β, whose orbit is a translation line L that is not closed in the plane. In fact, its closure contains the line $x = 1$. It is worth remarking that the line L does not separate the plane, but path-separates the plane into three distinct path components. In Figure 1d we illustrate a translation strip containing the point p and a translation strip containing the point q. The translation domains are, respectively, $x < 1$ and $x > -1$. Note also that h moves the translation strip for p "downward," even though h moves the iterates of the point s to points of successively greater height. Finally, the translation strip for p is transverse to the translation lines pictured in Figure 1a.

Note that in Figure 1c the point z lies both in the closed translation line $x = -1$ and the "wild" translation line L.

5. Example 2

In Figure 2, the homeomorphism h moves points of K (K is the union of the line $x = -1$ and the horizontal segments) directly down by distance one and moves points along the illustrated curves. (We shall indicate in the next paragraph a more formal method of defining the homeomorphisms in Figures 1 and 2.) The point v does not lie on a closed translation line. Hence [A] h cannot be the time one map of a topological flow.

The following proposition, whose proof is left to the reader, will allow us to formally define many of the homeomorphisms discussed in this paper.

PROPOSITION 3. *Let A be a closed subset of the plane. Then the map which moves each point of the plane horizontally to the right by $1/2$ its distance from A is a homeomorphism of the plane onto itself.*

Let $A = \{(x, y): x = -1 \text{ or } x = 1\}$, and let $S = \{(x, y): -1 \le x \le 1\}$. Let V be the "partial shear" homeomorphism defined by

$$V(x, y) = \begin{cases} (x, y - 1), & x \le -1, \\ (x, y + 1), & x \ge 1, \\ (x, y + x), & -1 \le x \le 1. \end{cases}$$

Let H be the homeomorphism that is the identity outside of S and which moves points of S horizontally to the right by $1/2$ their distance from A. The Reeb homeomorphism is conjugate to the composition $V \circ H$.

Let K be the union of the lines $x = \pm 1$ and the horizontal segments depicted in Figure 2, and let S be the strip $-1 \le x \le 1$. Let H be the homeomorphism which is the identity off of S and which moves points of S horizontally to the right by $1/2$ their distance from K. The homeomorphism of Figure 2 is conjugate to the composition $V \circ H$ where V is the partial shear described in the previous paragraph.

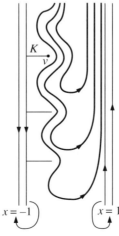

FIGURE 2

Thus, the examples in Figures 1 and 2 are both conjugate to "transvections," that is, finite compositions of homeomorphisms which are of the form $(x, y) \rightarrow (x, f(x, y))$ or $(x, y) \rightarrow (f(x, y), y)$. We say that the former type is a *vertical homeomorphism*, and the latter is a *horizontal homeomorphism*.

6. The theorem of Andrea

If h is a Brouwer homeomorphism and A is a subset of the plane, then A is defined to be "dissipative" provided that both the forward and backward orbits of A converge to infinity. Define an equivalence relation on the points of the plane by $x \sim y$ iff there is a dissipative arc containing both x and y. The equivalence classes of this equivalence relation are called the *fundamental regions* of h. The fundamental regions of h are obviously path connected and h permutes them, that is, if E is a fundamental region then so is $h(E)$. The fundamental regions are useful invariants with which to distinguish various conjugacy invariant properties of a (Brouwer) homeomorphism.

EXAMPLE. The homeomorphism of Figure 1 has three fundamental regions: $\{x \leq -1\}$, $\{x \geq 1\}$, and $\{-1 < x < 1\}$. The homeomorphism of Figure 2 also has three fundamental regions.

The following Propositions represent the principal results of Andrea's paper [A].

PROPOSITION 4 (S. Andrea). *A Brouwer homeomorphism cannot have exactly two fundamental regions.*

PROPOSITION 5 (S. Andrea). *A Brouwer homeomorphism with exactly one fundamental region is conjugate to a translation.*

Andrea observes that it is easy to construct examples of Brouwer homeomorphisms with any (positive) number of fundamental regions other than two, even a countable infinity.

Ault [AU] gives an example of a Brouwer homeomorphism with c fundamental regions, where c is the power of the continuum. A similar example occurs in [HT]. The example is constructed by modifying the Reeb homeomorphism. One inserts between each pair of a countably dense set of translation lines a third one so that the strip determined by the original two translation lines (see Figure 3a) is modified to look like Figure 3b.

In a paper presented to the Spring Topology Conference at Blacksburg in 1981, Utz [U] reviewed work of Andrea and Ault and drew further attention to the importance of fundamental regions in determining the flowability of Brouwer homeomorphisms. At that conference a number of questions concerning fundamental regions for Brouwer homeomorphisms were raised: (1) Is a fundamental region always unbounded? (2) Is a fundamental region always open or closed? (3) Is an open fundamental region always invariant? We answer all of these questions with the next few examples.

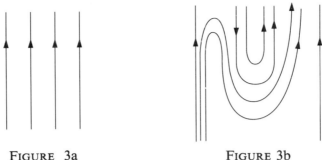

FIGURE 3a   FIGURE 3b

EXAMPLE 3. THE PLUNGER. (See Figure 4.) The homeomorphism h depicted in Figure 4 is invariant on the pictured translation lines and is invariant on the union of the $\sin 1/x$ curves. h moves each $\sin 1/x$ curve down to the one below. By a construction similar to the previous examples, one can show that h is conjugate to the composition of a horizontal homeomorphism with a vertical homeomorphism. Each $\sin 1/x$ curve together with the limit set of the previous one is a fundamental region. Thus h has fundamental regions that are bounded, and neither open, closed, nor invariant. This example could be easily modified by replacing the $\sin 1/x$ curve with another curve as long as the curve is path connected, does not separate the plane and the union of successive ones is not path connected.

EXAMPLE 4. (See Figure 5.) A homeomorphism with no closed invariant line.

This example is conceptually the same as the one constructed in [BST], but we describe it in a completely different way. The presentation given here has the advantage of being relatively simple to describe (the homeomorphism is a simple transvection), but the fundamental regions are more difficult to identify than in [BST].

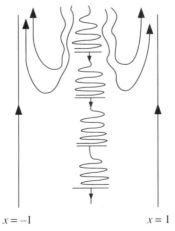

$x = -1$ $x = 1$

FIGURE 4

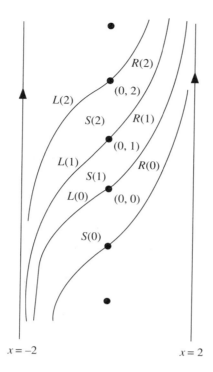

FIGURE 5

Let v be the vertical homeomorphism of the infinite strip $S = (-2 \leq x \leq 2)$ defined by

$$v(x, y) = (x, y + |x| - 1).$$

Let h be the homeomorphism of S that moves points horizontally to the right by $1/2$ their distance from the set consisting of the union of the vertical lines $x = -2$, $x = 2$, and the discrete set of points $((0, k)$, k an integer).

Let $g = h \circ v$. It is *not* obvious that the fundamental regions of this homeomorphism of S are of five types (see Figure 5):

(1) the lines $x = +2$, $x = -2$,
(2) the open rays $L(1)$, $L(2)$, ...,
(3) the open rays $R(1)$, $R(2)$, ...,
(4) the individual points $(0, n)$,
(5) the infinite open strip $S(n)$ each bounded above by $L(n) \cup \{(0, n)\} \cup R(n)$ and below by $L(n-1) \cup \{(0, n-1)\} \cup R(n-1)$.

The homeomorphism carries the fundamental region $L(n)$ to the fundamental region $L(n-1)$ and carries the fundamental region $R(n)$ to the fundamental region $R(n-1)$. Obviously $(0, n)$ is carried to $(0, n-1)$, and hence $S(n)$ is carried to $S(n-1)$. So the only invariant fundamental regions are the vertical lines $x = -2$, $x = 2$. We wish to construct a homeomorphism of the plane with no invariant fundamental region. We do this by "replacing" the vertical lines by strings of "plungers" as in Example 3. In the modified construction, the individual plungers are each noninvariant fundamental regions. By gluing together infinitely many copies of the "strip" and repeating the construction on each copy we get the homeomorphism of the plane with no invariant fundamental regions constructed in [BST].

We note that a necessary and sufficient condition that a Brouwer homeomorphism have no invariant closed line is that it has no invariant fundamental region, because an invariant fundamental region always contains an invariant closed line (see Prop. 3.2 of [A]). On the other hand, it is easily seen that an invariant closed line is a subset of an invariant fundamental region. Thus, our example, having no invariant fundamental region, has no invariant closed line.

QUESTION. Suppose h is a fixed point free orientation-preserving homeomorphism of the plane. Define an equivalence relation on the points of the plane by: x is equivalent to y provided that there is a *continuum* containing x and y that is dissipative under the iterates of h. Let us call the equivalence classes of this relation the generalized fundamental domains of h. Is there an h with no invariant generalized fundamental domain?

REFERENCES

[A] S. A. Andrea, *On homeomorphisms of the plane which have no fixed points*, Abh. Math. Sem. Univ. Hamburg **30** (1967), 61–74.

[AU] R. Ault, *Embedding of fixed-point free homeomorphisms of the plane in continuous flows*, Dissertation, Brandeis Univ., 1970.

[BRWR] L. E. J. Bouwer, *Beweis des ebenen Translationssatzes*, Math. Ann. **72** (1912), 37–54.

[B] M. Brown, *Homeomorphisms of two-dimensional manifolds*, Houston J. Math. **11** (1985), 455–469.

[BST] M. Brown, E. Slaminka, and W. Transue, *An orientation preserving fixed point free homeomorphism of the plane which admits no closed invariant line*, Topology **29** (1988), 213–217.

[HT] T. Homma and H. Terasaka, *On the structure of the plane translation theorem of Brouwer*, Osaka Math. J. **5** (1953), 233–266.

[K] Kerekjarto, *Vorlesungen uber Topologie*, Springer, Berlin, 1923.

[U] W. Utz, *The embeddings of homeomorphisms in continuous flows*, Topology Proc. **6** (1981), 159–175.

DEPARTMENT OF MATHEMATICS, UNIVERSITY OF MICHIGAN, ANN ARBOR, MICHIGAN 48109

Contemporary Mathematics
Volume **117**, 1991

Dense Orbits of Critical Points for the Tent Map

K. M. BRUCKS, B. DIAMOND,
M. V. OTERO-ESPINAR, AND C. TRESSER

In this note we work with the family of *tent maps* $f_s : [0, 2] \to [0, 2]$, $0 \leq s \leq 2$, defined by

$$f_s(x) = \begin{cases} sx, & 0 \leq x \leq 1, \\ s(2 - x), & 1 \leq x \leq 2. \end{cases}$$

We shall consider the set

$$\mathscr{D} = \left\{ s \in \left[\sqrt{2}, 2\right] \mid O_s(1) \text{ is dense in } I_s \right\},$$

where $O_s(1) = \{f_s^n(1) \mid n \in \mathbb{N}\}$ is the orbit of 1 under the map f_s, and where I_s stands for the invariant interval $[f_s^2(1), f_s(1)]$. We note that I_s is the core of this one-dimensional system, i.e., everything is attracted to this interval except 0 and 1. The purpose of this note is to provide, within these proceedings, an affirmative answer to a question raised by J. Martin during the Arcata Conference on Continuum Theory and Dynamical Systems (1989)—is the set $\mathscr{D}$ dense in $[\sqrt{2}, 2]$? In fact we prove more; namely that $\mathscr{D}$ is G_δ-dense in $[\sqrt{2}, 2]$.

A parameter $s \in [0, 2]$ is periodic if $f_s^m(1) = 1$ for some $m \in \mathbb{N}$, and is prefixed if the nonzero fixed point $z_s = 2s/(1 + s)$ of f_s is in the orbit of 1 under f_s. For each $n \in \mathbb{N}$, set $\varphi_n(s) = f_s^n(1)$, $s \in [0, 2]$. Our work heavily uses the density of the periodic and prefixed parameters in $[\sqrt{2}, 2]$. Both results can be found in [**2**], but for the convenience of the reader, we include new and simpler proofs of both facts.

LEMMA 1. *The periodic parameters are dense in* $[\sqrt{2}, 2]$.

PROOF. Suppose that there exists some $U = [\alpha, \beta] \subset [\sqrt{2}, 2]$ such that U contains no periodic parameters. Choose $n \in \mathbb{N}$ such that

$$(*) \qquad\qquad |\varphi_n'(s)| > 2/(\beta - \alpha)$$

1980 *Mathematics Subject Classification* (1985 *Revision*). Primary 34C35.
This paper is in final form and no version of it will be submitted for publication elsewhere.

for any $s \in U$ such that $\varphi'_n(s)$ exists. (The existence of such an n is precisely Lemma 5.3 of [1]. This is a computation depending only on the fact that $s \in [\sqrt{2}, 2]$.) Thus if $\varphi'_n(s)$ exists for all $s \in U$, then, by the Mean Value Theorem, there exists an $s_0 \in U$ such that

$$|\varphi'_n(s_0)| = \left| \frac{\varphi_n(\beta) - \varphi_n(\alpha)}{\beta - \alpha} \right|.$$

But since $|\varphi_n(\beta) - \varphi_n(\alpha)| \leq 2$, we have that $|\varphi'_n(s_0)| \leq 2/(\beta - \alpha)$. This contradicts $(*)$. Thus, there is some $t \in U$ such that $\varphi'_n(t)$ does not exist, i.e., $f_t^m(1) = 1$ for some $m < n$ so that t is periodic of period less than n. This completes the proof.

REMARK. In fact, Lemma 1 holds true in [1,2], [2, p. 235].

The original version of the following lemma appeared such a long time ago that we could not trace back its origin. The given proof follows the lines of [3, p. 78].

LEMMA 2. *Let* $s \in (\sqrt{2}, 2]$ *and* $J \subseteq [0, 2]$ *be an open interval. Then there is some* $n \in \mathbb{N}$ *so that* $I_s \subseteq f_s^n(J)$.

PROOF. For each positive integer i either (1) $|f_s^{i+1}(J)| \geq (s^2|f_s^{i-1}(J)|)/2$, or (2) $f_s^{i+1}(J) = I_s$. To see this, let K be some interval in $[0, 2]$. Then, $|f_s(K)| \geq s|K|$ if $1 \notin K$ and $|f_s(K)| \geq s|K|/2$ if $1 \in K$. Thus, $|f_s^2(K)| \geq s^2|K|/2$ if $1 \notin K \cap f_s(K)$. Also, $1 \in K \cap f_s(K)$ implies that 1, $f_s(1) \in f_s(K)$ and therefore that $f_s^2(K) \supset I_s$. Hence, (1) or (2) must hold.

Lastly, $s > 1$ implies that (1) cannot hold for all i, from which Lemma 2 follows.

For $\alpha \neq \beta$ in $\mathbf{R}$, let $\langle \alpha, \beta \rangle$ denote the interval with endpoints α, β.

LEMMA 3. *The prefixed parameters are dense in* $[\sqrt{2}, 2]$.

PROOF. Let $U \subset [\sqrt{2}, 2]$ be an open interval. Since the periodic parameters are dense we can choose $t \in U$ such that there exists an m with $\varphi_m(t) = 1$ and $\varphi_j(t) \neq 1$ for $0 < j < m$. Let $(j_1, j_2, \ldots, j_m)$ denote the permutation of the integers from 1 to m so that $\varphi_{j_1}(t) < \varphi_{j_2}(t) < \cdots < \varphi_{j_m}(t)$. (For example, if $m = 3$ and $\varphi_2(t) < \varphi_3(t) < \varphi_1(t)$, then $j_1 = 2$, $j_2 = 3$, and $j_3 = 1$. In fact, we always have that $j_1 = 2$ and $j_m = 1$.) Using Lemma 2 we have that the preimages of the fixed point $z_t = 2t/(1 + t)$ under the map f_t are dense in $[0, 2]$. Thus, for each i, $1 \leq i \leq m - 1$, let z_i be such that $\varphi_{j_i}(t) < z_i < \varphi_{j_{i+1}}(t)$, and such that z_i is a preimage of z_t under f_t. Since $f_s(x)$ varies continuously in x and s, there exists $\epsilon > 0$ and continuous functions $g_1, \ldots, g_{m-1}$ defined on $[t, t + \epsilon) \subset U$ such that for each i, $i \leq i \leq m - 1$, we have

   1) $g_i(t) = z_i$,
   2) for each $s \in [t, t + \epsilon)$, $g_i(s)$ is a preimage of $z_s = 2s/(1 + s)$ under f_s,
   3) $g_i(s) \neq g_j(s)$ for $i \neq j$ and $s \in [t, t + \epsilon)$.

Suppose that there do not exist prefixed parameters in $[t, t + \epsilon)$. Choose $t_1 \in (t, t + \epsilon)$ so that the kneading sequence of t_1 is aperiodic and not eventually periodic. (We can do this since the periodic and eventually periodic parameters are algebraic and therefore countable.) Set

$$\Delta_1 = [\varphi_2(t_1), g_1(t_1)], \Delta_2 = [g_1(t_1), g_2(t_1)], \ldots, \Delta_m = [g_{m-1}(t_1), \varphi(t_1)].$$

Note that $g_1(t_1) < g_2(t_1) < \cdots < g_m(t_1)$. Now, no prefixed parameters in $[t, t + \epsilon)$ imply that for every pair of nonnegative integers p, q and each i, $1 \le i \le m$, we have that any interval $W_{p,q,i}$ with endpoints of the form

$$\varphi_{pm+j_i}(t_1), \varphi_{qm+j_i}(t_1),$$

is such that $W_{p,q,i} \subset \Delta_r$, some r. (Note that if $\varphi_m(s) = g_i(s)$ for some s, m, i, then s is a prefixed parameter.) Let i_0 be such that $1 \in \Delta_{i_0}$. Then, for $i \ne i_0$, f_{t_1} is monotone on Δ_i, and thus $f_{t_1}(W_{p,q,i}) = \langle \varphi_{pm+j_r}(t_1), \varphi_{qm+j_r}(t_1) \rangle$ is contained in Δ_r, for some r. For $i = i_0$, f_{t_1} may not be monotone on W_{p,q,i_0}. However, $f_{t_1}(W_{p,q,i_0}) = \langle \varphi_{(p+1)m+1}(t_1), \varphi_{(q+1)m+1}(t_1) \rangle \subset \Delta_m$ or $f_{t_1}(W_{p,q,i_0}) = \langle \min\{\varphi_{(p+1)m+1}(t_1), \varphi_{(q+1)m+1}(t_1)\}, \varphi_1(t_1) \rangle \subset \Delta_m$. Thus, if $k_1 \ne k_2$ are positive integers, then the nondegenerate interval G with endpoints $\varphi_{k_1 m + 2}(t_1), \varphi_{k_2 m + 2}(t_1)$ has the property that for each $n \ge 0$:

$$f_{t_1}^n(G) \subset \Delta_{i_n}, \quad \text{some } i_n \in \{1, \ldots, m\}.$$

Hence $I_s \not\subseteq f_{t_1}^n(G)$ for all n, contradicting Lemma 2. See Figure 1.

LEMMA 4. *For any open interval V in $(\sqrt{2}, 2)$, there is an integer n so that*

$$I_{\sqrt{2}} \subseteq \varphi_n(V) \cup \varphi_{n+1}(V).$$

PROOF. Choose a periodic parameter s_0 in V and a prefixed parameter t in V with $t > s_0$, say $f_{s_0}^m(1) = 1$ and $f_t^j(1) = 2t/(1 + t)$. Choose k so that $km > j$. Then $\varphi_{km+1}(s_0) = s_0 = f_{s_0}(1)$ while $\varphi_{km+1}(t) = 2t/(1 + t)$, and $\varphi_{km+2}(s_0) = f_{s_0}^2(1)$ while $\varphi_{km+2}(t) = f_t^2(1)$. That is,

$$[2t/(1+t), f_{s_0}(1)] \subseteq \varphi_{km+1}(V) \quad \text{and} \quad [f_{s_0}^2(1), 2t/(1+t)] \subseteq \varphi_{km+2}(V).$$

The lemma now follows since $I_{\sqrt{2}} \subseteq I_{s_0}$.

Finally we get

THEOREM 5. *The set $\mathscr{D}$ is G_δ-dense in $[\sqrt{2}, 2]$.*

PROOF. For each rational number $r \in I_{\sqrt{2}}$ and each $n \in \mathbb{N}$, define

$$U_{r, 1/n} = \left\{ s \in \left(\sqrt{2}, 2\right) \mid O_s(1) \cap (r - 1/n, r + 1/n) \ne 0 \right\}.$$

Since $\varphi_m(s)$ is a continuous function of s for each $m \in \mathbb{N}$, the set $U_{r, 1/n}$ is open. It follows from Lemma 4 that $U_{r, 1/n}$ is dense in $(\sqrt{2}, 2]$. According

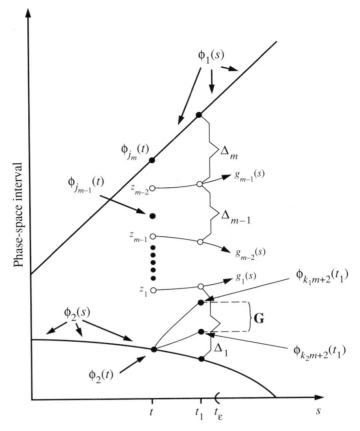

FIGURE 1. Construction of G

to Lemma 2, if the orbit of 1 under f_s is dense in $I_{\sqrt{2}}$, then it is also dense in I_s, $s \in [\sqrt{2}, 2]$.

Hence,

$$\mathscr{D} = \bigcap_{r \in I_{\sqrt{2}}} \bigcap_{n \in \mathbb{N}} U_{r, 1/n}.$$

As a countable intersection of dense open subsets of $[\sqrt{2}, 2]$, the set $\mathscr{D}$ is G_δ-dense in $[\sqrt{2}, 2]$. This completes the proof of Theorem 5.

The theory of the period doubling renormalization group is trivially solved for tent maps [1]. Using it, Theorem 5 immediately implies that the set $\mathscr{D}'$ of parameter values for which the orbit of 1 is dense in the core is G_δ-dense in [1,2]. (In general, if $\sqrt{2} < s^m \le 2$, $m = 1, 2, 2^2, 2^3, \ldots$, then the core of f_s consists of m disjoint intervals and a finite number of periodic points [3, p. 78].) This will be detailed in a coming paper which will *contain* a more complete study of the topological dynamics of the tent family.

REFERENCES

1. P. Coullet and C. Tresser, *Iterations d'endormorphismes et groupe de renormalisation*, C. R. Acad. Sci. Paris, **287** (1978), 577.
2. E. M. Coven, I. Kan, and J. A. Yorke, *Pseudo-orbit shadowing in the family of tent maps*, Trans. Amer. Math. Soc. **308** (1988), 227–241.
3. S. van Strien, *Smooth dynamics on the interval*, New Directions in Dynamical Systems, LMS Lecture Notes, vol. 127, Cambridge Univ. Press, New York, 1988, pp. 57–119.

INSTITUTE FOR MATHEMATICAL SCIENCES, SUNY AT STONY BROOK, STONY BROOK, NEW YORK, 1794-3660

DEPARTMENT OF MATHEMATICS, COLLEGE OF CHARLESTON, CHARLESTON, SC 29424

DEPT. DE ANALISE MATEMATICA, CAMPUS UNIVERSITARIO, S.N. 15771-SANTIAGO DE COMPOSTELA, SPAIN

IBM RESEARCH DIVISION, T. J. WATSON RESEARCH CENTER, YORKTOWN HEIGHTS, NY 10598

Contemporary Mathematics
Volume 117, 1991

Periods of Surface Homeomorphisms

JOHN FRANKS AND JAUME LLIBRE

ABSTRACT. The goal of this paper is to investigate which sets of positive integers can occur as the periods of the periodic orbits of a surface homeomorphism on a given compact surface. We also investigate the influence of the induced map on homology on the sets of periods which can occur.

1. Introduction and statement of results

Compact connected 2-dimensional manifolds are called *surfaces*. Any orientable surface without boundary is homeomorphic to the sphere S^2 or to the torus T^2 or to the connected sum of n tori with $n \geq 2$ (i.e. the n-holed torus). The *genus* of an orientable surface without boundary is the number of torus summands.

Let f be a surface homeomorphism. We denote by $\operatorname{Per}(f)$ the set of periods of all periodic points of f.

Fuller, in [**Fu**], proved the following result; see also Halpern [**Hl**] and Brown [**Br**].

THEOREM 1. *Let f be a homeomorphism of a compact polyhedron X into itself. If the Euler characteristic of M is not zero, then f has a periodic point with period not greater than the maximum of $\sum_{k \, odd} B_k(X)$ and $\sum_{k \, even} B_k(X)$, where $B_k(X)$ denotes the k-th Betti number of X.*

If we apply Theorem 1 to surface homeomorphisms we obtain

COROLLARY 2. *Let S be an orientable surface without boundary of genus g and let $f : S \to S$ be a homeomorphism. Then the following statements hold:*

(1) *If $g = 0$ then $\operatorname{Per}(f) \cap \{1, 2\} \neq \varnothing$.*
(2) *If $g > 1$ then $\operatorname{Per}(f) \cap \{1, 2, \ldots, 2g\} \neq \varnothing$.*

PROOF. It is well known that $B_0(S) = B_2(S) = 1$ and that $B_1(S) = 2g$

1980 *Mathematics Subject Classification* (1985 *Revision*). Primary 58F20.
This paper is in final form and no version of it will be submitted for publication elsewhere.

(for more details, see [Ca]). Then the Euler characteristic of S is $\chi(S) = 2(1 - g)$. Therefore $\chi(S) \neq 0$ if and only if $g \neq 1$. Hence the corollary follows immediately from Theorem 1. $\square$

To improve Corollary 2 we need some notation. Let A be an $n \times n$ complex matrix. A $k \times k$ *principal submatrix* of A is a submatrix lying in the same set of k rows and columns, and a $k \times k$ *principal minor* is the determinant of such a principal submatrix. There are $\binom{n}{k}$ different $k \times k$ principal minors of A, and the sum of these is denoted by $E_k(A)$. In particular, $E_1(A)$ is the trace of A, and $E_n(A)$ is the determinant of A, denoted by $\det(A)$.

If $\lambda_1, \ldots, \lambda_n$ are the eigenvalues of A, then it is well known that the characteristic polynomial of A satisfies

$$
\begin{aligned}
\det(tI - A) &= t^n - E_1(A)t^{n-1} + E_2(A)t^{n-2} - \cdots + (-1)^n E_n(A) \\
&= t^n - S_1(\lambda_1, \ldots, \lambda_n)t^{n-1} + S_2(\lambda_1, \ldots, \lambda_n)t^{n-2} \\
&\quad - \cdots + (-1)^n S_n(\lambda_1, \ldots, \lambda_n),
\end{aligned}
$$

where $S_k(\lambda_1, \ldots, \lambda_n)$ is the kth *elementary symmetric function* of the n numbers $\lambda_1, \ldots, \lambda_n$ for $k \leq n$, defined as follows:

$$
S_k(\lambda_1, \ldots, \lambda_n) = \sum_{1 \leq i_1 < \cdots < i_k \leq n} \prod_{j=1}^{k} \lambda_{i_j},
$$

i.e. the sum of all $\binom{n}{k}$ k-fold products of distinct items from $\lambda_1, \ldots, \lambda_n$. Thus $S_1(\lambda_1, \ldots, \lambda_n) = \lambda_1 + \cdots + \lambda_n$ and $S_n(\lambda_1, \ldots, \lambda_n) = \lambda_1 \cdots \lambda_n$.

The next two theorems improve Corollary 2. They deal with orientation-reversing and orientation-preserving homeomorphisms of surfaces, respectively.

THEOREM 3. *Let S be an orientable surface without boundary of genus g and let $f : S \to S$ be an orientation-reversing homeomorphism. Suppose that A is the $2g \times 2g$ integral matrix of the isomorphism $f_{*1} : H_1(S; \mathbb{Q}) \to H_1(S; \mathbb{Q})$ induced by f on the first rational homology group. Then the following statements hold*:

(1) *If $g = 0$ then $\mathrm{Per}(f) \cap \{1, 2\} \neq \varnothing$.*
(2) *If $g > 0$ and $E_1(A) \neq 0$, then $1 \in \mathrm{Per}(f)$.*
(3) *If $g > 0$, $E_1(A) = 0$ and $E_2(A) \neq -1$, then $\mathrm{Per}(f) \cap \{1, 2\} \neq \varnothing$.*
(4) *If $g = 1$, $E_1(A) = 0$ and $E_2(A) = -1$, then there are no restrictions for the set $\mathrm{Per}(f)$.*
(5) *If $g > 1$, $E_1(A) = 0$, $E_2(A) = -1$ and k is the smallest integer of the set $\{3, 4, \ldots, 2g\}$ such that $E_k(A) \neq 0$, then f has a periodic point of period a divisor of k.*

THEOREM 4. *Let S be an orientable surface without boundary of genus g and let $f : S \to S$ be an orientation-preserving homeomorphism. Suppose* *that A is the $2g \times 2g$ integral matrix of the isomorphism $f_{*1} : H_1(S; \mathbb{Q}) \rightarrow H_1(S; \mathbb{Q})$ induced by f on the first rational homology group. Then the following statements hold*:

(1) *If $g = 0$ then $1 \in \text{Per}(f)$.*

(2) *If $g > 0$ and $E_1(A) \neq 2$, then $1 \in \text{Per}(f)$.*

(3) *If $g > 0$, $E_1(A) = 2$ and $E_2(A) \neq 1$, then $\text{Per}(f) \cap \{1, 2\} \neq \varnothing$.*

(4) *If $g = 1$, $E_1(A) = 2$ and $E_2(A) = 1$, then there are no restrictions for the set $\text{Per}(f)$.*

(5) *If $g > 1$, $E_1(A) = 2$, $E_2(A) = 1$ and k is the smallest integer of the set $\{3, 4, \dots, 2g\}$ such that $E_k(A) \neq 0$, then f has a periodic point of period a divisor of k.*

Theorems 3 and 4 will be proved in §2.

Similar results to those given in Theorems 3 and 4 are given in [**Ll**] for continuous self-maps of a connected finite graph.

Let f be a surface homeomorphism and let A be a $2g \times 2g$ integral matrix of the isomorphism f_{*1} induced by f on the first rational homology group. A condition on the eigenvalues of A is given by the following result.

PROPOSITION 5. *Let S be an orientable surface without boundary of genus g and let $f : S \rightarrow S$ be a homeomorphism. Suppose that A is the $2g \times 2g$ integral matrix of the isomorphism $f_{*1} : H_1(S; \mathbb{Q}) \rightarrow H_1(S; \mathbb{Q})$ induced by f on the first rational homology group. Then the following statements hold*:

(1) *If f is orientation-reversing and λ is an eigenvalue of A of multiplicity k, then $\bar{\lambda}$, $-\lambda^{-1}$, $-\bar{\lambda}^{-1}$ are eigenvalues of A of multiplicity k ($\bar{\lambda}$ denotes the complex conjugate of λ).*

(2) *If f is orientation-preserving and λ is an eigenvalue of A of multiplicity k, then $\bar{\lambda}$, λ^{-1}, $\bar{\lambda}^{-1}$ are eigenvalues of A of multiplicity k. Moreover, the multiplicities of the eigenvalues $+1$ and -1, if they occur, are even.*

Proposition 5 is well known, but case (1) does not appear to exist in the literature. Since the proof is easy, we give it in §3.

We denote by $\text{Diff}_-(S)$ the space of all orientation-reversing C^1 diffeomorphisms of the orientable surface without boundary S. A natural question is the following: *Assuming that $f \in \text{Diff}_-(S)$ and that $\text{Per}(f)$ is finite, what integers must be elements of $\text{Per}(f)$?*

Of course the analogous question for orientation-preserving surface diffeomorphisms is uninteresting because it is easy to construct orientation-preserving diffeomorphisms on any orientable surface with fixed points and an arbitrary finite set of higher periods.

The following theorems give a partial answer to the above question. Most of what is known concerning necessary conditions which $\text{Per}(f)$ must satisfy is due to results from [**BF**] in the genus zero case and [**Hn**] in the positive genus case. In the following three theorems we summarize the results of these papers which apply to this question and add sufficiency results and a new necessary condition in the case of genus 2 or 3. The three theorems apply to the sphere, the torus, and the higher genus case, respectively. We denote by $2\mathbb{N}$ the set of all even positive integers.

THEOREM 6. *The following statements hold*:

(1) *If* $f \in \text{Diff}_-(S^2)$ *has finitely many periods and finitely many fixed points, then* $\text{Per}(f)$ *is a finite subset of* $2\mathbb{N} \cup \{p\}$ *where* p *is an odd positive integer, and* $\text{Per}(f)$ *satisfies either* $\text{Per}(f) = \{1\}$ *or* $2 \in \text{Per}(f)$.

(2) *If* A *is any finite subset of* $2\mathbb{N} \cup \{p\}$ *where* p *is an odd positive integer, and* A *satisfies either* $A = \{1\}$ *or* $2 \in A$, *then there exists* $f \in \text{Diff}_-(S^2)$ *such that* $\text{Per}(f) = A$.

THEOREM 7. *The following statements hold*:

(1) *If* $f \in \text{Diff}_-(T^2)$ *has finitely many periods, then* $\text{Per}(f)$ *is a finite subset of* $2\mathbb{N} \cup \{p, q\}$ *where* p *and* q *are distinct odd positive integers.*

(2) *If* A *is any finite subset of* $2\mathbb{N} \cup \{p, q\}$ *where* p *and* q *are odd positive integers, then there exists* $f \in \text{Diff}_-(T^2)$ *such that* $\text{Per}(f) = A$.

THEOREM 8. *Let* S *be an orientable surface without boundary of genus* $g > 1$.

(1) *If* $f \in \text{Diff}_-(S)$ *has finitely many periods, then* $\text{Per}(f)$ *is a finite subset of* $2\mathbb{N} \cup \{p_1, \dots, p_{g+1}\}$ *where* $p_1, \dots, p_{g+1}$ *are distinct odd positive integers, and* $\text{Per}(f) \cap \{1, \dots, 2g\} \neq \varnothing$. *Furthermore, if* $g \in \{2, 3\}$ *then* $\text{Per}(f) \cap \{1, 2, 4\} \neq \varnothing$.

(2) *If* A *is any finite subset of* $2\mathbb{N} \cup \{p_1, \dots, p_{g+1}\}$ *where* $p_1, \dots, p_{g+1}$ *are distinct odd positive integers, and* $A \cap \{1, 2\} \neq \varnothing$, *then there exists* $f \in \text{Diff}_-(S)$ *such that* $\text{Per}(f) = A$.

Assertion (1) of Theorems 6, 7, and 8 is proved in §4, using a result of [Hn], and Assertion (2) of Theorems 6 and 7 is proved in §5. Finally, Theorem 8(2) is showed in §6.

The proofs of Assertion (2) of Theorems 6 and 7 use the characterization of orientation-reversing Morse-Smale diffeomorphisms of the sphere S^2 and of the torus T^2 given by Batterson, Handel, and Narasimhan [BHN] and by Batterson [B2], respectively. An alternative proof can be done by using the classification of braid types for periodic orbits of orientation-reversing diffeomorphisms of surfaces of genus 0 and 1 with topological entropy zero due to Llibre and MacKay [LM2] and to Guaschi, Llibre, and MacKay [GLM], respectively (see [W] also in this regard).

Note that we give a complete answer to our question for orientation-reversing surface diffeomorphisms of the torus, a nearly complete answer for the sphere because our result needs that f has finitely many fixed points, and only a partial answer for orientable surfaces without boundary of genus greater than one. More information on this last case is given in §§5 and 6.

Finally, in §7 we study the set of periods for orientation-reversing diffeomorphisms on orientable surfaces with boundary.

An interesting work also related to the periods of surface homeomorphisms with zero topological entropy, but from another point of view, is due to Smillie [**Smi**].

We would like to thank Robert MacKay for his comments on a preliminary version of this paper, and to Nonlinear Systems Laboratory of the Mathematics Institute of the University of Warwick for making these conversatons possible by sponsoring the Warwick Reunion on Braid Types of periodic orbits for surface homeomorphisms. The second author is partially supported by a DGICYT grant, no. PB86-0351.

2. Periods and homology

In this section we shall prove Theorem 3 by using the Lefschetz fixed point theorem and the Lefschetz zeta function. The proof of Theorem 4 is similar.

Let S be an orientable surface without boundary of genus g, and let $f : S \to S$ be an orientation-reversing homeomorphism. The map f induces isomorphisms $f_{*k} : H_k(S; \mathbb{Q}) \to H_k(S; \mathbb{Q})$ (for $k = 0, 1, 2$) on the rational homology groups of $S (H_0(S; \mathbb{Q}) \approx \mathbb{Q}, H_1(S; \mathbb{Q}) \approx \mathbb{Q} \oplus \overset{2g}{\cdots} \oplus \mathbb{Q}, H_2(S; \mathbb{Q}) \approx \mathbb{Q})$ such that $f_{*0} = \mathrm{id}$, $f_{*2} = -\mathrm{id}$, and f_{*1} is given by a $2g \times 2g$ integral matrix A, where $2g$ is the rank of $H_1(S; \mathbb{Q})$ (for more details see [**Ms**]).

The *Lefschetz number* of f is defined by

$$L(f) = \mathrm{trace}(f_{*0}) - \mathrm{trace}(f_{*1}) + \mathrm{trace}(f_{*2}).$$

For our homeomorphism f of S the Lefschetz fixed point theorem states (see [**Br**]):

THEOREM 9. *If $L(f) \neq 0$ then f has a fixed point.*

When studying the periodic points of f it is convenient to use the *Lefschetz zeta function* of f defined by

(a) $$Z_f(t) = \frac{\det(I - t f_{*1})}{\det(I - t f_{*0})\det(I - t f_{*2})} = \frac{\det(I - tA)}{1 - t^2}.$$

If A is the 0×0 matrix then take $\det(I - tA) = 1$. It can be viewed as a rational function of t or, expanding the denominator, as a power series in t. In fact $Z_f(t)$ is a generating function for the sequence of Lefschetz numbers $L(f^n)$ since

(b) $$Z_f(t) = \exp\left(\sum_{n=1}^{\infty} \frac{L(f^n)}{n} t^n\right).$$

This equality follows easily taking into account that

$$L(f^n) = \begin{cases} -\mathrm{trace}(A^n) & \text{if } n \text{ is odd,} \\ 2 - \mathrm{trace}(A^n) & \text{if } n \text{ is even,} \end{cases}$$

$\log(1 - t^2) = -\sum_{n=1}^{\infty} t^{2n}/n$, and

$$\frac{1}{\det(I - tA)} = \exp\left(\sum_{n=1}^{\infty} \frac{\text{trace}(tA)^n}{n}\right).$$

For more details on the Lefschetz zeta function see [**Fr**].

If $g = 0$ then from (a) we have $Z_f(t) = (1 - t^2)^{-1}$. Therefore, from (b) it follows that $L(f) = 0$ and $L(f^2) = 2$. Hence, from Theorem 9, assertion (1) of Theorem 3 follows.

If $g > 0$ and $E_1(A) \neq 0$, then from (a) and (b) it follows that $L(f) = -E_1(A)$. Hence, from Theorem 9, assertion (2) of Theorem 3 is proved.

If $g > 0$, $E_1(A) = 0$, and $E_2(A) \neq -1$, then from (a) and (b) we obtain that $L(f) = 0$ and $L(f^2) = 2[1 + E_2(A)]$. Hence, from Theorem 9, assertion (3) of Theorem 3 holds.

If $g = 1$, $E_1(A) = 0$, and $E_2(A) = -1$, then from (a) we obtain $Z_f(t) = 1$. From (b) it follows that $L(f^n) = 0$ for $n = 1, 2, \ldots$. So, Theorem 9 does not give any restriction on the set $\text{Per}(f)$.

Furthermore, given any set of positive integers P it is possible to construct a homeomorphism f of the torus whose set of periods is precisely P. If P is the empty set, any irrational translation of the torus will do. We sketch the construction when P is infinite, since the finite case is similar and easier. Let $\{q_n\}_{n=1}^{\infty}$ be the elements of P arranged in increasing order and let α be any positive irrational number. Choose a sequence of positive integers $\{p_n\}_{n=1}^{\infty}$ such that

$$\lim p_n/q_n = \alpha.$$

On the torus choose a sequence of disjoint parallel circles $\{C_n\}$ which limit on an essential circle D. If we rotate each C_n by p_n/q_n and D by α, then this map can be extended to a homeomorphism of the torus with no other periodic points. The orbit of any point between two of the parallel circles will be asymptotic to one of the circles in forward time and to the other in backward time. This constructions allows any set P of positive integers to be realized as $\text{Per}(f)$. Thus assertion (4) of Theorem 3 is proved.

Finally, assume that $g > 1$, $E_1(A) = 0$, $E_2(A) = -1$, and k is the smallest integer of the set $\{3, 4, \ldots, 2g\}$ such that $E_k(A) \neq 0$. Since f_{*1} is an isomorphism, $\det(A) = E_{2g}(A) \neq 0$. So k is well defined. From (a) we have

$$Z_f(t) = \frac{1 - t^2 + (-1)^k E_k(A)t^k + \cdots + (-1)^{2g} E_{2g}(A)t^{2g}}{1 - t^2}$$

$$= 1 + (-1)^k E_k(A)t^k + O(t^{k+1}).$$

Then, from (b) we obtain that

$$L(f) = L(f^2) = \cdots = L(f^{k-1}) = 0,$$

and
$$L(f^k) = (-1)^k k E_k(A).$$

Therefore, from Theorem 9, f^k has a fixed point and assertion (5) of Theorem 3 is proved.

3. Eigenvalues of the first homology isomorphism

In this section we shall prove assertion (1) of Proposition 5; assertion (2) is a well-known property of symplectic automorphisms and follows in a similar way.

Let S be an orientable surface without boundary of genus g and let $f: S \to S$ be an orientation-reversing homeomorphism. The matrix $J = \begin{pmatrix} 0 & -I \\ I & 0 \end{pmatrix}$ is the $2g \times 2g$ matrix for the non singular intersection pairing $H_1(S; \mathbb{Q}) \times H_1(S; \mathbb{Q}) \to \mathbb{Q}$, i.e. this pairing is given by $\langle x, Jy \rangle$, where $\langle , \rangle$ is a standard inner product with respect to some basis. Suppose that A is the $2g \times 2g$ integral matrix of the isomorphism $f_{*1}: H_1(S; \mathbb{Q}) \to H_1(S; \mathbb{Q})$ induced by f with respect to this basis. The fact that f reverses orientation implies that $\langle Ax, JAy \rangle = -\langle x, Jy \rangle$ for all $x, y \in H_1(S; \mathbb{Q})$. It follows that $A^t J A = -J$, so

$$\begin{aligned} \det(I - tA) &= \det(J^t J - tA) = \det(-J^t A^t J A - tA) \\ &= \det(A)\det(-J^t A^t J - tI) = \det(A)\det(-J^t A^t J - tJ^t J) \\ &= \det(A)\det(-A^t - tI) = \det(A)t^{2g}\det(-t^{-1}A^t - I) \\ &= \det(A)t^{2g}\det(I + t^{-1}A^t) = \det(A)t^{2g}\det(I + t^{-1}A). \end{aligned}$$

Since $\det(A) \neq 0$, A has no zero eigenvalues. So if λ is an eigenvalue of A then $-\lambda^{-1}$ also is an eigenvalue of A. Now, assertion (1) of Proposition 5 follows taking into account that A is a real matrix.

4. Restrictions to the set of periods

In this section we shall show that the set of periods of an orientation-reversing surface diffeomorphism f with finitely many periods have some restrictions, first due to the topology of the surface (see, for instance, Theorems 3 and 4) and then due to the finiteness of $\operatorname{Per}(f)$.

From Theorem 3(1) it follows that if $f \in \operatorname{Diff}_-(S^2)$ then $\operatorname{Per}(f) \cap \{1, 2\} \neq \varnothing$. Now we shall see that if f is an orientation-reversing diffeomorphism with finitely many periods and finitely many fixed points; then we obtain more restrictions for its set of periods. The next proposition was proved for orientation-reversing Morse-Smale diffeomorphisms of S^2 by Batterson, Handel, and Narasimhan [BHN], but as we shall show that their proof (with small changes) works for any orientation-reversing diffeomorphism of S^2 with finitely many periods and finitely many fixed points.

PROPOSITION 10. *Let $f \in \operatorname{Diff}_-(S^2)$ and assume that f has finitely many fixed points. Then either $\operatorname{Per}(f) = \{1\}$ or $2 \in \operatorname{Per}(f)$.*

PROOF. Suppose that $\mathrm{Per}(f) \neq \{1\}$ and that $2 \notin \mathrm{Per}(f)$. By Theorem 3(1), $\mathrm{Per}(f) \cap \{1, 2\} \neq \varnothing$. So $1 \in \mathrm{Per}(f)$ and if $n \in \mathrm{Per}(f)$ with $n \neq 1$ then $n > 2$. Now we proceed in a similar way to the proof of Proposition 3 of [BHN]. The first step is to delete from S^2 all the fixed points and an orbit of period $n > 2$. The resulting space is then recompactified to a new surface N by adding a boundary circle for each deleted point. The diffeomorphism f is extended to a homeomorphism g of N by a radial projection of the derivative. Note that $\mathrm{Per}(g|_{N \setminus \partial N}) \cap \{1, 2\} = \varnothing$. As in the proof of Proposition 3 of [BHN], we obtain that there exists a g^2-Nielsen class of index 1.

On each boundary circle of N all elements of $\mathrm{Fix}\,(g^2)$ are in the same g^2-Nielsen class. Since these fixed points of g^2 can be removed by a homotopy, they do not contribute to the g^2-fixed point index. Consequently, we have a contradiction with the existence of a g^2-Nielsen class of index 1. □

For orientation-reversing surface diffeomorphisms with finitely many periods we can obtain restrictions to the set of periods from the next theorem, which was conjectured by Blanchard and Franks [BF] (who also gave a proof when the genus is 0) and proved by Handel [Hn].

THEOREM 11. *Let S be an orientable surface without boundary of genus g. If $f \in \mathrm{Diff}_-(S)$ has periodic points with $g + 2$ distinct odd periods, then $\mathrm{Per}(f)$ is infinite.*

From Corollary 2, Proposition 10, and Theorem 11, assertion (1) of Theorems 6, 7, and 8 follows with the exception of the claim in Theorem 8(1) that if the genus g is 2 or 3 then 1, 2, or 4 is a period of the orientation-reversing diffeomorphism. To prove this claim we shall need the following auxiliary results.

The following theorem is due to Manning [Mn].

THEOREM 12. *Let S be an orientable surface without boundary of genus g and let $f : S \to S$ be a homeomorphism. Suppose that A is the $2g \times 2g$ integral matrix of the isomorphism $f_{*1} : H_1(S; \mathbb{Q}) \to H_1(S; \mathbb{Q})$ induced by f on the first rational homology group. If A has an eigenvalue λ with $|\lambda| > 1$, then the topological entropy $h(f)$ of f is positive.*

The following result follows from the Thurston's classification of surface homeomorphisms; for more details see [Hn].

THEOREM 13. *Let S be an orientable surface without boundary. Any C^1 diffeomorphism $f : S \to S$ with positive topological entropy has periodic orbits with infinitely many distinct periods.*

From Proposition 5 and Theorems 12 and 13, the next result follows easily.

COROLLARY 14. *Let S be an orientable surface without boundary of genus g and let $f : S \to S$ be a C^1 diffeomorphism. Suppose that A is the $2g \times 2g$*

integral matrix of the isomorphism $f_{*1} : H_1(S; \mathbb{Q}) \to H_1(S; \mathbb{Q})$ induced by f on the first rational homology group. If $\mathrm{Per}(f)$ is finite then all the eigenvalues of A have modulus equal to the unity.

From Proposition 5 and Corollary 14 we obtain easily the following result.

COROLLARY 15. Let S be an orientable surface without boundary of genus g and suppose that $f \in \mathrm{Diff}_-(S)$ has finitely many periodic points. Assume that A is the $2g \times 2g$ integral matrix of the isomorphism $f_{*1} : H_1(S; \mathbb{Q}) \to H_1(S; \mathbb{Q})$ induced by f on the first rational homology group. Then the characteristic polynomial of A is a product of factors of the following forms, $t^2 + 1$, $t^2 - 1$, and $t^4 - 2\cos(2\alpha)t^2 + 1$, where α is the real part of an eigenvalue of A.

Now the remainder part of assertion (1) of Theorem 8 follows from the next proposition.

PROPOSITION 16. Let S be an orientable surface without boundary of genus g and suppose that $f \in \mathrm{Diff}_-(S)$ has finitely many periodic points. If $g \in \{2, 3\}$ then $\mathrm{Per}(f) \cap \{1, 2, 4\} \neq \varnothing$.

PROOF. Let A be the $2g \times 2g$ integral matrix of the isomorphism $f_{*1} : H_1(S; \mathbb{Q}) \to H_1(S; \mathbb{Q})$ induced by f on the first rational homology group.

First suppose that $g = 2$. Since $\det(tI - A)$ is a polynomial in t of degree 4, from Corollary 15 it must be equal to one of the following four polynomials: $(t^2 + 1)^2$, $(t^2 + 1)(t^2 - 1)$, $(t^2 - 1)^2$, $t^4 - 2\cos(2\alpha)t^2 + 1$ where $2\cos(2\alpha)$ is an integer. Then, from Theorem 3 it follows that $\mathrm{Per}(f) \cap \{1, 2\} \neq \varnothing$ except if $\det(tI - A) = t^4 - t^2 + 1$. But in this exceptional case Theorem 3(5) states that $\mathrm{Per}(f) \cap \{1, 2, 4\} \neq \varnothing$.

Now assume that $g = 3$. Again from Corollary 15, $\det(tI - A)$ must be equal to one of the following six polynomials: $(t^2 + 1)^3$, $(t^2 + 1)^2(t^2 - 1)$, $(t^2 + 1)(t^4 - 2\cos(2\alpha)t^2 + 1)$, $(t^2 - 1)(t^4 - 2\cos(2\alpha)t^2 + 1)$, $(t^2 - 1)^2(t^2 + 1)$, $(t^2 - 1)^3$. Then from Theorem 3 it follows that $\mathrm{Per}(f) \cap \{1, 2\} \neq \varnothing$ except if $\det(tI - A)$ is either $(t^2 - 1)^2(t^2 + 1)$ or $(t^2 - 1)(t^4 + 1)$. But in these two exceptional cases Theorem 3(5) states that $\mathrm{Per}(f) \cap \{1, 2, 4\} \neq \varnothing$. $\square$

We note that using Theorem 3 it is possible to improve in Theorem 8(1) the condition $\mathrm{Per}(f) \cap \{1, \ldots, 2g\} \neq \varnothing$ for $g > 3$ in a way similar to cases $g = 2$ and $g = 3$.

5. Orientation-reversing Morse-Smale diffeomorphisms

In this section we shall prove Theorems 6(2) and 7(2) by using the classification of the orientation-reversing Morse-Smale diffeomorphisms of S^2 and T^2. We begin by recalling several definitions. Further details may be found in [Fr] and [Sma].

Let f be a diffeomorphism of a compact manifold M whose nonwandering set consists of a finite number of hyperbolic orbits. Then to each periodic orbit γ of f, we may associate the triple (p, u, Δ), where p is the period of $x \in \gamma$, u (the *index* of γ) is the dimension of E_x^u for $x \in \gamma$, and Δ (the *orientation type* of γ) is $+1$ if $Df_x^p : E_x^u \to E_x^u$ preserves orientation and -1 if it reverses orientation. The collection of such triples $\{(p_i, u_i, \Delta_i)\}_{i=1}^n$, where n is the number of periodic orbits of f, is called the *periodic data* of f. Note that the same triple may occur more than once if it is associated to more than one orbit.

In the case of f with finitely many periodic orbits, all hyperbolic, Franks has shown that the homology zeta function has a simple form in terms of the periodic data of f.

THEOREM 17. *Let $f : M \to M$ be a diffeomorphism of a compact manifold M with periodic data $\{(p_i, u_i, \Delta_i)\}_{i=1}^n$. Then the Lefschetz zeta function of f is*

$$Z_f(t) = \prod_{i=1}^n (1 - \Delta_i t^{p_i})^{(-1)^{u_i+1}} .$$

Note that the denominator product of $Z_f(t)$ is over the sink and source periodic orbits and that the numerator product ranges over the saddle periodic orbits.

A *Morse-Smale diffeomorphism* is a diffeomorphism f of M satisfying the following conditions:

(1) the set of nonwandering points $\Omega(f)$ is finite;
(2) the periodic points of f are hyperbolic;
(3) for each $x, y \in \Omega(f)$, the stable manifold of x and the unstable manifold of y have transversal intersection.

Condition (1) implies that $\Omega(f)$ consists of periodic points. Palis and Smale proved in [PS] that for a diffeomorphism f of M with $\Omega(f)$ finite, f is a Morse-Smale diffeomorphism if and only if f is structurally stable. For this reason, it is of interest to know what Morse-Smale diffeomorphisms exist on a given manifold. The next proposition summarizes some necessary conditions to have an orientation-reversing Morse-Smale diffeomorphism of a surface without boundary.

PROPOSITION 18. *If there exists an orientation-reversing Morse-Smale diffeomorphism f of a surface without boundary of genus g with periodic data $\{(p_i, u_i, \Delta_i)\}_{i=1}^n$, then*

(1) $u_i = 0$ *and* $u_j = 2$ *for some i and j*;
(2) $\Delta_i = 1$ *if* $u_i = 0$ *and* $\Delta_i = (-1)^{p_i}$ *if* $u_i = 2$;
(3) $Z_f(t) = \det(I - At)/(1 - t^2) = \prod_{i=1}^n (1 - \Delta_i t^{p_i})^{(-1)^{u_i+1}}$, *where A is the matrix associated to the isomorphism f_{*1}*;
(4) $\{p_i\}_{i=1}^n$ *contains at most $g + 1$ different odd numbers.*

PROOF. Condition (1) is simply the statement that a Morse-Smale diffeomorphism of a compact manifold must have at least one source and one sink.

Condition (2) is the statement that the orientation type of sinks is $+1$, while the orientation type of the unstable manifold of a source is preserved by an even iterate of an orientation-reversing map, and reversed by an odd iterate of such a map.

Condition (3) follows from the definition of the Lefschetz zeta function and Theorem 17.

Finally, Condition (4) follows from Theorem 11. □

Note that to compute $\det(I - At)$ in Proposition 18(3) it suffices to calculate A for the isotopy class of f.

The following two theorems of [BHN] and [B2], respectively, characterize the orientation-reversing Morse-Smale diffeomorphisms of S^2 and of T^2.

THEOREM 19. *There exists an orientation-reversing Morse-Smale diffeomorphism f of S^2 with periodic data $\{(p_i, u_i, \Delta_i)\}_{i=1}^{n}$ if and only if*

(1) $u_i = 0$ *and* $u_j = 2$ *for some* i *and* j;

(2) $\Delta_i = 1$ *if* $u_i = 0$ *and* $\Delta_i = (-1)^{p_i}$ *if* $u_i = 2$;

(3) $Z_f(t) = (1 - t^2)^{-1}$;

(4) $\{p_i\}_{i=1}^{n}$ *contains at most one different odd number*;

(5) *If* $p_i > 2$, *for some* i, *then the data contains one of the following triples*: $(2, 0, 1)$, $(2, 2, 1)$, *or* $(2, 2, -1)$.

In [B1] it was proved that to study the periodic behavior of Morse-Smale orientation-reversing diffeomorphisms of the torus, it suffices to consider the isotopy classes of the toral diffeomorphisms induced by $\left(\begin{smallmatrix} 1 & 0 \\ 0 & -1 \end{smallmatrix}\right)$ and $\left(\begin{smallmatrix} 1 & 1 \\ 0 & -1 \end{smallmatrix}\right)$. A toral diffeomorphism is of *even type* if it belongs to the isotopy class induced by $\left(\begin{smallmatrix} 1 & 0 \\ 0 & -1 \end{smallmatrix}\right)$. Those sharing this relationship with $\left(\begin{smallmatrix} 1 & 1 \\ 0 & -1 \end{smallmatrix}\right)$ are of *odd type*. If p is odd we say that the collection $\{(p, 0, 1), (p, 2, -1), (2p, 1, 1)\}$ is a *minimal data collection*.

THEOREM 20. *There exists an orientation-reversing Morse-Smale diffeomorphism f of T^2 with periodic data $\{(p_i, u_i, \Delta_i)\}_{i=1}^{n}$ if and only if*

(1) $u_i = 0$ *and* $u_j = 2$ *for some* i *and* j;

(2) $\Delta_i = 1$ *if* $u_i = 0$ *and* $\Delta_i = (-1)^{p_i}$ *if* $u_i = 2$;

(3) $Z_f(t) = 1$;

(4) $\{p_i\}_{i=1}^{n}$ *contains at most two different odd numbers*;

(5) *If* f *is of even type then* $\{(p_i, u_i, \Delta_i)\}_{i=1}^{n}$ *is not a minimal data collection*.

The following proposition is a direct consequence of Lemma 2.1 of [Na] and Lemma 5 of [B1]. It may be used to organize the set of periods of an orientation-reversing Morse-Smale diffeomorphism of an orientable surface without boundary.

PROPOSITION 21. *Let* $\{(p_i, u_i, \Delta_i)\}_{i=1}^n$ *be a collection of triples satisfying the first four condition of Proposition 18; then the product*

$$(Z_f(t))^{-1} \cdot \prod_{i=1}^n (1 - \Delta_i t^{p_i})^{(-1)^{u_i+1}} = 1$$

can be expressed without cancellation as

$$\prod_{i=1}^{m_1} \left(\frac{1 + t^{p_i}}{1 + t^{p_i}}\right)^{r_i} \cdot \left[\frac{1 - t^{2p_i}}{(1 - t^{p_i})(1 + t^{p_i})}\right]^{s_i}$$

$$\cdot \prod_{j=1}^{m_2} \frac{(1 - t^{q_j})(1 + t^{q_j})(1 + t^{2q_j}) \cdots (1 + t^{2^{n_j}q_j})}{(1 - t^{2^{n_j}q_j})},$$

where for each i, p_i *is odd,* $r_i \geq 0$, *and* $s_i \geq 0$; *for each* j, *either* q_j *is equal to some* p_i *or* q_j *is even; if* $n_j = 0$ *then the corresponding term is* $(1 - t^{q_j})/(1 - t^{q_j})$; $m_k \geq 0$, $m_1 \leq g+1$, *and* $m_1 + m_2 > 0$. *If some* $m_k = 0$ *then the product associated to* m_k *is* 1.

From Theorems 17 and 19 and Proposition 21, Theorem 6(2) follows. Similarly, from Theorems 17 and 20 and Proposition 21, Theorem 7(2) follows.

6. Orientation-reversing diffeomorphisms of surfaces of genus greater than one with periods 1 or 2

In this section we shall prove Theorem 8(2).

PROOF OF THEOREM 8(2) WHEN 1 BELONGS TO A. From Theorems 17 and 20 and Proposition 21, there exist f_1 and f_2 in $\mathrm{Diff}_-(T^2)$ such that $\mathrm{Per}(f_1)$ and $\mathrm{Per}(f_2)$ have at most one arbitrary odd positive integer different from 1 and a number of arbitrary even positive integers. Furthermore, f_1 has two fixed points with triples $(1, 1, 1)$ and $(1, 0, 1)$ associated to a factor $(1 - t)/(1 - t)$ of $Z_{f_1}(t)$, and f_2 has two fixed points with triples $(1, 1, 1)$ and $(1, 2, 1)$ associated to a factor $(1 - t)/(1 - t)$ of $Z_{f_2}(t)$. Of course, a fixed point with triple $(1, 0, 1)$ (resp. $(1, 2, 1)$) is a sink (resp. a source).

Let p be the fixed point with triple $(1, 0, 1)$ for f_1 and let q be the fixed point with triple $(1, 2, 1)$ for f_2. We choose a small topological disk D_p on a torus T_p^2 around p in such a way that there is a neighborhood U_p of ∂D_p in T_p^2 such that $f(U_p) \subset D_p$. Similarly we consider a small topological disk D_q on a torus T_q^2 (different from T_p^2) around q in such a way that there is a neighborhood U_q of ∂D_q in T_q^2 such that $f^{-1}(U_q) \subset D_q$.

Now we remove the interiors of D_p and D_q from T_p^2 and T_q^2, respectively, and we consider the connected sum of $T_p^2 \setminus \mathrm{Int}(D_p)$ with $T_q^2 \setminus \mathrm{Int}(D_q)$ with respect to ∂D_p and ∂D_q in such a way that f_1 and f_2 extend to an orientation-reversing diffeomorphism f on the 2-holed torus S thus obtained. Then, by construction of f it follows that $\mathrm{Per}(f)$ has the period 1, at most two arbitrary odd positive integers different from 1, and a number of arbitrary even positive integers. Hence Theorem 8(2) is proved for $g = 2$ when $1 \in A$. If $g > 2$, we make a connected sum between the 2-holed torus S with a removed fixed point of triple $(1, 0, 1)$, and a torus T^2 with a removed fixed point of triple $(1, 2, 1)$, to obtain an orientation-reversing diffeomorphism of an orientable surface of genus 3 satisfying Theorem 8(1) when $1 \in A$, and so on. $\square$

PROOF OF THEOREM 8(2) WHEN 2 BELONGS TO A. From Theorems 17 and 20 and Proposition 21, there exists $f_1 \in \text{Diff}_-(T^2)$ such that $\text{Per}(f_1)$ has at most two arbitrary odd positive integers, a number of arbitrary even positive integers, and two periodic orbits with triples $(2, 1, 1)$ and $(2, 0, 1)$ associated to a factor $(1 - t^2)/(1 - t^2)$ of $Z_{f_1}(t)$. Similarly, from Theorems 17 and 19 and Proposition 21, there exists $f_2 \in \text{Diff}_-(S^2)$ such that $\text{Per}(f_2)$ has at most one arbitrary odd positive integer, a number of arbitrary even positive integers, and one periodic orbit with triples $(2, 2, 1)$ associated to a factor $1/(1 - t^2)$ of $Z_{f_2}(t)$.

Let $\{p_1, p_2\}$ be the periodic orbit with triple $(2, 0, 1)$ for f_1 and let $\{q_1, q_2\}$ be the periodic orbit with triple $(2, 2, 1)$ for f_2. We choose two small topological disks D_{p_1} and D_{p_2} on T^2 around p_1 and p_2, respectively, in such a way that there is a neighborhood U_{p_i} of ∂D_{p_i} in T^2 such that $f(U_{p_i}) \subset D_{p_i}$. Similarly we consider two small topological disks D_{q_1} and D_{q_2} on S^2 around q_1 and q_2, respectively, in such a way that there is a neighborhood U_{q_i} of ∂D_{q_i} in S^2 such that $f^{-1}(U_{q_i}) \subset D_{q_i}$.

Now we remove the interiors of D_{p_i} and D_{q_i} from T^2 and S^2, respectively, and we consider the connected sum of $T^2 \setminus (\bigcup_{i=1}^2 \text{Int}(D_{p_i}))$ with $S^2 \setminus (\bigcup_{i=1}^2 \text{Int}(D_{q_i}))$ with respect to ∂D_{p_i} and ∂D_{q_i} in such a way that f_1 and f_2 extend to an orientation-reversing diffeomorphism f on the 2-holed torus S thus obtained. Then, by construction of f it follows that $\text{Per}(f)$ has at most three arbitrary odd positive integers, a number of arbitrary even positive integers, and the period 2. Hence Theorem 8(2) is proved for $g = 2$ when $2 \in A$. If $g > 2$, we make a connected sum between the 2-holed torus S with a removed periodic orbit of triple $(2, 0, 1)$, and a sphere S^2 with a removed periodic orbit of triple $(2, 2, 1)$, to obtain an orientation-reversing diffeomorphism of an orientable surface of genus 3 satisfying Theorem 8(2) when $2 \in A$, and so on. $\square$

We note that Theorem 8 completes the study of the set of periods for an orientation-reversing diffeomorphism f on an orientable surface of genus 2 (resp. 3), except in the case that $Z_f(t) = (1 - t^2 + t^4)/(1 - t^2)$ (resp. $1 \pm t^4$) and $\text{Per}(f) \cap \{1, 2\} = \varnothing$ (see the proof of Proposition 16). Of course, in these exceptional cases we know that $4 \in \text{Per}(f)$. It is easy to construct diffeomorphisms satisfying the above exceptional cases.

7. Orientation-reversing diffeomorphisms of orientable surfaces with boundary

Any compact orientable surface with boundary is homeomorphic to an orientable surface without boundary with a finite number of small disks removed. By the *genus* of an orientable surface with boundary S, we understand the genus of the associated orientable surface without boundary obtained by gluing on a finite number of disks to the boundary of S. More details on orientable surfaces may be found in [Ca] and [Ko].

Let S be an orientable surface with boundary. The boundary of S, ∂S, is formed by a set of disjoint simple closed curves denoted by B_k for $k = 1, \dots, r$ and called the *boundary components* of S. If $f \in \mathrm{Diff}_-(S)$ then ∂S is invariant by f and its boundary components may be permuted by f. We say that B_k has period n if $f^n(B_k) = B_k$ and $n \geq 1$ is the least such integer. We define $P(f)$ as the set of all positive integers n such that f has a periodic point of period n, or f has a boundary component of period n.

The study of the set $P(f)$ for an orientation-reversing diffeomorphism f of an orientable surface S with boundary of genus g, will be reduced to analyzing the set of periods of an orientation-reversing diffeomorphism F of S_g called the *completion* of f, where S_g denotes the orientable surface without boundary of genus g.

Think of S as a subset of S_g with all its boundary components circles of the same radius a. We define the completion $F : S_g \to S_g$ of $f : S \to S$ as follows (see [Ep] or [LM1]). Let $F = f$ on S. If C is a boundary component of S and D_C, the disk which it bounds, we extend the diffeomorphism $f|_C : C \to f(C)$ to a diffeomorphism from D_C to $D_{f(C)}$ by $F(r, \theta) = (r - \varepsilon(r, \theta), f|_C(\theta))$, using polar coordinates in each disk. Here $\varepsilon(r, \theta) > 0$ for $0 < r < a$ and $\varepsilon(r, \theta) = 0$ for $r = 0$ and $r = a$. Therefore the center of the disk D_C is a periodic point of the same period that C for f. Note that the center of the disk D_C is the unique periodic point of F in the interior of D_C. Hence $P(f) = \mathrm{Per}(F)$. Now to study $P(f)$ we can apply the results obtained in the previous sections for $\mathrm{Per}(F)$.

The completion defined here is not exactly the completion of [Ep] or [LM1], but in this way we have that if f has finitely many periodic points then its completion F also has finitely many periodic points.

REFERENCES

[B1]   S. Batterson, *The dynamics of Morse-Smale diffeomorphisms on the torus*, Trans. Amer. Math. Soc. **256** (1979), 395–403.

[B2]   ———, *Orientation reversing Morse-Smale diffeomorphisms on the torus*, Trans. Amer. Math. Soc. **264** (1981), 29–37.

[BHN] S. Batterson, M. Handel, and C. Narasimhan, *Orientation reversing Morse-Smale diffeomorphism of S^2*, Invent. Math. **64** (1981), 345–356.

[BF]   P. Blanchard and J. Franks, *The dynamical complexity of orientation reversing homeomorphisms of surfaces*, Invent. Math. **62** (1980), 333–339.

[Br]   R. F. Brown, *The Lefschetz fixed point theorem*, Scott-Foresman, Chicago, 1971.

[Ca] S. S. Cairns, *Introductory topology*, The Ronald Press Company, New York, 1968.

[Ep] D. B. A. Epstein, *Pointwise periodic homeomorphisms*, Proc. London Math. Soc. **42** (1981), 415–460.

[Fr] J. Franks, *Homology and dynamical systems*, CBMS Regional Conf. Series, vol. 49, Amer. Math. Soc., Providence, R. I., 1982.

[Fu] F. B. Fuller, *The existence of periodic points*, Ann. of Math. **57** (1953), 229–230.

[GLM] J. Guaschi, J. Llibre, and R. S. MacKay, *A classification of braid types for periodic orbits of diffeomorphisms of surfaces of genus one with topological entropy zero*, J. London Math. Soc. (to appear).

[Hl] B. Halpern, *Fixed points for iterates*, Pacific J. Math. **25** (1968), 255–275.

[Hn] M. Handel, *The entropy of orientation reversing homeomorphisms of surfaces*, Topology **21** (1982), 291–296.

[Ko] C. Kosniowski, *A first course in algebraic topology*, Cambridge Univ. Press, New York, 1980.

[Ll] J. Llibre, *Periodic points of one dimensional maps*, European Conference on Iteration Theory - 1989, World Scientific, Singapore, 1990 (to appear).

[LM1] J. Llibre and R. S. MacKay, *Rotation vectors and entropy for homeomorphisms of the torus isotopic to the identity*, Ergodic Theory Dynamical Systems (to appear).

[LM2] _____, *A classification of braid types for diffeomorphisms of surfaces of genus zero with topological entropy zero*, J. London Math. Soc. (to appear).

[Mn] A. Manning, *Topological entropy and the first homology group*, Dynamical Systems - Warwick 1974, Lecture Notes in Math., vol. 468, Springer-Verlag, Berlin, 1975, pp. 185–199.

[Ms] W. S. Massey, *Singular homology theory*, Springer-Verlag, New York, 1980.

[Na] C. Narasimhan, *The periodic behavior of Morse-Smale diffeomorphisms on compact surfaces*, Trans. Amer. Math. Soc. **248** (1979), 145–169.

[PS] J. Palis and S. Smale, *Structural stability theorems*, Proc. Sympos. Pure Math., vol. 14, Amer. Math. Soc., Providence, R. I., 1970, pp. 223–231.

[Sma] S. Smale, *Differentiable dynamical systems*, Bull. Amer. Math. Soc. **73** (1967), 747–817.

[Smi] J. Smillie, *Periodic points of surface homeomorphisms with zero entropy*, Ergodic Theory Dynamical Systems **3** (1983), 315–334.

[Th] W. P. Thurston, *On the geometry and dynamics of diffeomorphisms of surfaces*, Bull. Amer. Math. Soc. **19** (1988), 417–431.

[W] M. Wada, *Closed orbits of non-singular Morse-Smale flows*, J. Math. Soc. Japan **41** (1989), 405–413.

DEPARTMENT OF MATHEMATICS, NORTHWESTERN UNIVERSITY, EVANSTON, ILLINOIS 60208, USA

E-mail address: john@math.nwu.edu

DEPARTAMENT DE MATEMÀTIQUES, UNIVERSITAT AUTÒNOMA DE BARCELONA, BELLATERRA, 08193 BARCELONA, SPAIN

Contemporary Mathematics
Volume 117, 1991

Fixed-point Problems in Continuum Theory

CHARLES L. HAGOPIAN

ABSTRACT. L. E. J. Brouwer published his paper containing the Brouwer fixed-point theorem two years after he constructed an indecomposable plane continuum. For twenty-six years, these two results appeared to be unrelated. Today they provide the foundation for a major investigation in continuum theory. This investigation is centered on the following classical unsolved problem: Does every plane continuum that does not separate the plane have the fixed-point property? We give a brief history of this problem and discuss some related results and open questions.

A space M has the *fixed-point property* if for each map f of M into M there is a point x of M such that $f(x) = x$.

A *continuum* is a nondegenerate compact connected metric space.

The Brouwer fixed-point theorem [**Br**] asserts that every disk has the fixed-point property. Note that a disk is a plane continuum that does not separate the plane. Furthermore, every nonseparating plane continuum is the intersection of a nested sequence of (topological) disks. Hence an affirmative answer to the following question would provide a natural generalization to Brouwer's theorem.

QUESTION 1. Does every nonseparating plane continuum have the fixed-point property?

Question 1 is often called the plane fixed-point problem. According to R. H. Bing [**Bi3**, p. 122], it is one of the most interesting problems in plane topology. The plane fixed-point problem has motivated the development of a variety of concepts in continuum theory [**Bi4, BM, H2, Ma**]. It first appeared in a paper of W. L. Ayers [**Ay**] in 1930. Ayers [**Ay**] proved that every homeomorphism of a locally connected nonseparating plane continuum into itself has a fixed point. In 1932, K. Borsuk [**Bo**] introduced the concept of a

---

1980 *Mathematics Subject Classification* (1985 *Revision*). Primary 54F20, 54H25; Secondary 57N05.

Key words and phrases. Fixed-point property, nonseparating plane continuum, indecomposability, Brouwer fixed-point theorem.

This paper is in final form and no version of it will be submitted for publication elsewhere.

retract to generalize Ayers's theorem. Borsuk [**Bo**] proved that every locally connected nonseparating plane continuum has the fixed-point property by showing that every continuum of this type is a retract of a disk. In 1959, Borsuk and J. Stallings [**St**] were following this approach to Question 1 when they asked if every nonseparating plane continuum is an almost continuous retract of a disk.

A function f of a space M into a space N is *almost continuous* if for each open set G in $M \times N$ that contains f there exists a map g of M into N such that G contains g.

Since every almost continuous retract of a disk has the fixed-point property, an affirmative answer to their question would have answered Question 1 in the affirmative. However, V. Akis [**A1**] showed that the plane continuum formed by a disk and a ray spiraling to its boundary is not an almost continuous retract of a disk. Although the answer to Borsuk and Stallings's question is no, almost continuous retracts have been used to estabish the fixed-point property for several general classes of continua [**A2**].

A continuum is *indecomposable* if it is not the union of two proper subcontinua.

Brouwer discovered indecomposable continua in 1910 [**K1**, p. 204; **K2**, p. 72]. The first relationship between these continua and the fixed-point property was established by O. H. Hamilton [**Ha1**] in 1938. In 1967 and 1968, H. Bell [**B1**] and K. Sieklucki [**S**] independently generalized Hamilton's work by proving that every nonseparating plane continuum that admits a fixed-point-free map has an invariant indecomposable continuum in its boundary. This theorem was also established independently by S. D. Iliadis [**I**] in 1970. Recently P. Minc [**Mi3**] established the fixed-point property for every nonseparating plane continuum in which each indecomposable continuum in its boundary is contained in a weakly chainable continuum (i.e., a continuous image of a pseudo-arc).

A continuum M is *arc-like* (*star-like*, *tree-like*, or *disk-like*) if for each positive number ε, there is an ε-map of M onto an arc (star, tree, or disk, respectively). Recall that a tree is a finite graph that does not contain a simple closed curve. A star is a tree with only one junction point. Every arc-like continuum is star-like, every star-like continuum is tree-like, and every tree-like continuum is disk-like.

Bing [**Bi1**] proved that every arc-like continuum is a nonseparating plane continuum. In [**Ha2**], Hamilton proved that every arc-like continuum has the fixed-point property.

Since every tree-like plane continuum is a nonseparating plane continuum, the following open question provides another approach to the plane fixed-point problem.

QUESTION 2. Does every tree-like plane continuum have the fixed-point property?

In 1979, D. P. Bellamy [**Be**] constructed a nonplanar tree-like continuum that admits a fixed-point-free map. He defined this continuum by modifying a solenoid. Bellamy [**Be**, p. 12] used this example and an inverse limit technique of J. B. Fugate and L. Mohler [**FM**] to construct a second tree-like continuum that admits a fixed-point-free homeomorphism. It is not known if either of Bellamy's examples are star-like.

QUESTION 3. Does every star-like continuum have the fixed-point property?

It is not known if Bellamy's second example can be embedded in the plane [**OR1**, **OR2**]. Note that if it can, the answer to Question 1 is no. This example is an indecomposable continuum with the property that every proper subcontinuum is an arc. One might be able to show that it is not planar by answering the following question in the affirmative:

QUESTION 4. If M is a tree-like plane continuum with the propery that every proper subcontinuum is an arc, must M have the fixed-point property?

The behavior of a tree-like plane continuum is extremely unpredictable. Therefore it is reasonable to focus on general subclasses. For example, an answer to the following question would be considered a breakthrough.

QUESTION 5. Does every triod-like plane continuum have the fixed-point property?

Suppose M is a triod-like continuum that can be represented as the inverse limit of triods with bonding maps that fix every point of two arms of each triod. M. M. Marsh [**Mh1**, **Mh2**] has proved a general theorem that implies M has the fixed-point property.

QUESTION 6. If the product of two continua is disk-like, must it have the fixed-point property?

E. Dyer [**D**] proved that all products of arc-like continua have the fixed-point property. Hence if one could show that the factors in Question 6 are arc-like, it would follow that the answer to Question 6 is yes. Some partial solutions to this problem are given in [**H3**, **H4**, **H5**].

In [**Bi3**], Bing asked the following:

QUESTION 7. If M is a plane continuum with the fixed-point property, must $M \times [0, 1]$ have the fixed-point property?

Question 7 remains open when it is also assumed that M does not separate the plane. In fact, this question is open for arcwise connected nonseparating plane continua.

In 1971, the author [**H1**] used the Bell-Sieklucki theorem to prove that

(1)   every arcwise connected nonseparating plane continuum has the fixed-point property.

R. Bennett [**Ben**] proved that every locally connected disk-like continuum is planar. Hence every locally connected disk-like continuum has the fixed-point property.

Since there exist simple examples of disk-like arcwise connected continua that cannot be embedded in the plane, it is natural to ask the following:

QUESTION 8. Does every disk-like arcwise connected continuum have the fixed-point property?

A set is *uniquely arcwise connected* if it is arcwise connected and does not contain a simple closed curve.

In 1975, Mohler [**Mo**] used the Markov-Kakutani theorem to prove that every homeomorphism of a uniquely arcwise connected continuum into itself has a fixed point.

In 1976, the author [**H7**] proved that

(2)   every uniquely arcwise connected plane continuum has the fixed-point property.

These results answered two questions of Bing [**Bi3**, pp. 124, 126] and G. S. Young [**Y**]. The author's proof of (2) was based on the following intuitive concept of Bing [**Bi3**, p. 123–125]:

To prove that a certain uniquely arcwise connected continuum M has the fixed-point property, Bing assumed that M admitted a fixed-point-free map f. He imagined there was a dog at some point p of M and there was a rabbit at $f(p)$. The dog started chasing the rabbit and the rabbit naturally tried to escape. At every instant during the chase, the dog was advancing along the unique arc between himself and rabbit. The rabbit's position at every instant was the image of the dog's position under f. If M actually did admit such a map f, the dog would chase the rabbit forever in M without catching him. Bing used the structure of M to show that the rabbit would eventually get caught. Hence M did not admit a fixed-point-free map.

A map f of a space M is a *deformation* of M if there exists a map h of $M \times [0, 1]$ onto M such that $h(p, 0) = p$ and $h(p, 1) = f(p)$ for each point p of M.

According to the Lefschetz fixed-point theorem [**L**], every deformation of a polyhedron with nonzero Euler characteristic has a fixed point. The application of Lefschetz fixed-point theory to the entire class of uniquely arcwise connected continua is prohibited by the requirement that homology be finitely generated [**Mi1**, **Mi2**]. However, in 1984, the author [**H9**] used Bing's dog-chases-rabbit principle to prove that every deformation of a uniquely arcwise connected continuum has a fixed point.

Examples of uniquely arcwise connected continua that admit fixed-point-free maps have been given by Bing [**Bi3**], W. Holsztynski [**Ho**], Mohler and L. G. Oversteegen [**MO**], and Young [**Y**].

QUESTION 9. If M is a uniquely arcwise connected continuum and f is a fixed-point-free map of M into M, must M have a proper subcontinuum that is invariant under f?

In 1976, R. Manka [**M**] proved that every hereditarily unicoherent continuum that does not have an indecomposable subcontinuum has the fixed-point property. The double Warsaw circle shows that Manka's theorem cannot be extended to every hereditarily decomposable continuum that does not contain a simple closed curve.

QUESTION 10. Must every deformation of a hereditarily decomposable continuum that does not contain a simple closed curve have a fixed point?

In [**Le**, Problem 29], Bellamy asked:

QUESTION 11. Must every deformation of a tree-like continuum have a fixed point?

Although Question 11 is still open, the analogous question for nonseparating plane continua has been answered.

A map f of a space M is an *arc-component-preserving map* if f sends each arc-component of M into itself.

In [**H10**], the author proved that

(3)   every arc-component-preserving map of a nonseparating plane continuum has a fixed point.

Note that (3) generalizes (1).

Suppose M is a plane continuum, $\mathscr{D}$ is a decomposition of M, and each element of $\mathscr{D}$ is a uniquely arcwise connected set. In [**H12**], the author proved that every map of M that preserves the elements of $\mathscr{D}$ has a fixed-point. This result generalized (2) since it implies that every arc-component-preserving map of a plane continuum that does not contain a simple closed curve has a fixed point. This result also implies the Poincare-Bendixson theorem [**HS**, p. 248].

Every deformation is an arc-component-preserving map.

Hence we obtain a partial answer to Question 11 by noting that (3) implies

(4)   every deformation of a tree-like plane continuum has a fixed point.

R. L. Moore [**Mr**] proved that the plane does not contain uncountably many disjoint triods.

In [**H10**], the author generalized (4) by proving that

(5)   every arc-component-preserving map of a tree-like continuum that does not contain uncountably many disjoint triods has a fixed point.

It follows from (5) that neither of Bellamy's examples admit a fixed-point-free arc-component-preserving map.

QUESTION 12. Does every tree-like continuum have the fixed-point property for arc-component-preserving maps?

In [**H11**], the author took a small step toward the answers to Questions 11 and 12 by proving that every isotopic deformation of a tree-like continuum has a fixed-point.

The author proved (3) and (5) by applying Bing's dog-chases-rabbit principle to a free chain. The concept of a free chain had been introduced earlier to study homogeneous continua [**Bi2**, **H6**, **H8**]. Intuitively, a free chain can be thought of as a long, narrow, open-ended, hollow log through which the dog and rabbit run. This approach seems appropriate for Question 4. A continuum that has only arcs for proper subcontinua has an abundance of free chains. The dog should always be able to catch the rabbit in one of them.

Suppose M is a nonseparating plane continuum, f is a map of M into M, and only countably many arc-components of M are not preserved by f. Then the proof of (3) can be modified to show that f has a fixed point.

QUESTION 13. If M is a nonseparating plane continuum and f is a map of M into M that sends one arc-component of M into itself, must f have a fixed point?

The fixed-point-free map on Bellamy's first tree-like continuum sends one arc-component into itself. Bellamy's second example does not appear to have this property.

If Bellamy's second example can be embedded in the plane, the answer to the following long outstanding question [Ay] is no.

QUESTION 14. If M is a nonseparating plane continuum and h is a homeomorphism of M into M, must h have a fixed point?

The Cartwright-Littlewood theorem [CL] states that the answer to Question 14 is yes if it is also assumed that h extends to an orientation preserving homeomorphism of the plane. In 1976, Bell [B2] generalized the Cartwright-Littlewood theorem to every homeomorphism that extends to the plane.

REFERENCES

[A1] V. N. Akis, *Fixed point theorems and almost continuity*, Fund. Math. **121** (1984), 133–142.

[A2] ——, *Quasi-retractions and the fixed point property*, Contemp. Math. **72** (1988), 1–10.

[Ay] W. L. Ayres, *Some generalizations of the Scherer fixed-point theorem*, Fund. Math. **16** (1930), 332-336.

[B1] H. Bell, *On fixed point property of plane continua*, Trans. Amer. Math. Soc. **128** (1967), 539–548.

[B2] ——, *A fixed point theorem for plane homeomorphisms*, Fund. Math. **100** (1978), 119–128. (See, also, Bull. Amer. Math. Soc. **82** (1976), 778–780.)

[Be] D. P. Bellamy, *A tree-like continuum without the fixed point property*, Houston J. Math. **6** (1979), 1–13.

[Ben] R. Bennett, *Locally connected 2-cell and 2-sphere like continua*, Proc. Amer. Math. Soc. **17** (1966), 678–681.

[Bi1] R. H. Bing, *Snake-like continua*, Duke Math. J. **18** (1951), 653–663.

[Bi2] ——, *A simple closed curve is the only homogeneous bounded plane continuum that contains an arc*, Canad. J. Math. **12** (1960), 209–230.

[Bi3] ——, *The elusive fixed point property*, Amer. Math. Monthly **76** (1969), 119–132.

[Bi4] ——, *Commentary on Problem 107*, The Scottish Book, Birkhauser, Boston, 1981, pp. 190–192.

[Bo] K. Borsuk, *Einige Satze uber stetige Streckenbilder*, Fund. Math. **18** (1932), 198–213.

[BM] B. L. Brechner and J. C. Mayer, *The prime end structure of indecomposable continua and the fixed point property*, General Topology and Modern Analysis, Academic Press, NY, 1982, pp. 151–168.

[Br] L. E. J. Brouwer, *Uber Abbildung von Mannigfaltigkeiten*, Math. Ann. **71** (1912), 97–115.

[CL] M. L. Cartwright and J. E. Littlewood, *Some fixed point theorems*, Ann. of Math. **54** (1951), 1–37.

[D] E. Dyer, *A fixed point theorem*, Proc. Amer. Math. Soc. **7** (1956), 662–672.

[FM] J. B. Fugate and L. Mohler, *A note on fixed points in tree-like continua*, Topology Proc. **2** (1977), 457–460.

[H1] C. L. Hagopian, *A fixed point theorem for plane continua*, Bull. Amer. Math. Soc. **77** (1971), 351–354; Addendum in Bull. Amer. Math. Soc. **78** (1972), 289.

[H2] ____, *λ connectivity in the plane*, Studies in Topology, Academic Press, New York, 1975, pp. 197–202.

[H3] ____, *Disk-like products of λ connected continua. I*, Proc. Amer. Math. Soc. **51** (1975), 448–452.

[H4] ____, *Disk-like products of λ connected continua. II*, Proc. Amer. Math. Soc. **52** (1975), 479–484.

[H5] ____, *A fixed-point theorem for hyperspaces of λ connected continua*, Proc. Amer. Math. Soc. **53** (1975), 231–234.

[H6] ____, *A characterization of solenoids*, Pacific J. Math. **68** (1977), 425–435.

[H7] ____, *Uniquely arcwise connected plane continua have the fixed-point property*, Trans. Amer. Math. Soc. **248** (1979), 85–104.

[H8] ____, *Atriodic homogeneous continua*, Pacific J. Math. **113** (1984), 333–347.

[H9] ____, *The fixed-point property for deformations of uniquely arcwise connected continua*, Topology Appl. **24** (1986), 207–212.

[H10] ____, *Fixed points of arc-component-preserving maps*, Trans. Amer. Math. Soc. **306** (1988), 411–420.

[H11] ____, *Fixed points of tree-like continua*, Contemp. Math. **72** (1988), 131–137.

[H12] ____, *Fixed points of plane continua*, preprint.

[Ha1] O. H. Hamilton, *Fixed points under transformations of continua which are not connected im kleinen*, Trans. Amer. Math. Soc. **44** (1938), 18–24.

[Ha2] ____, *A fixed point theorem for pseudo-arcs and certain other metric continua*, Proc. Amer. Math. Soc. **2** (1951), 173–174.

[Ho] W. Holsztynski, *Fixed points of arcwise connected spaces*, Fund. Math. **64** (1969), 289–312.

[HS] M. W. Hirsch and S. Smale, *Differential equations, dynamical systems, and linear algebra*, Academic Press, New York, 1974.

[I] S. Iliadis, *Positions of continua on the plane and fixed points*, Vestnik Moskov. Univ. Ser. I Math. Mekh. (1970), 66–70.

[K1] K. Kuratowski, *Topology, vol. 2*, Academic Press, New York, 1968.

[K2] ____, *A half century of Polish mathematics*, International Series in Pure and Applied Mathematics, vol. 108, Pergamon Press, New York, 1980.

[L] S. Lefschetz, *Continuous transformations of manifolds*, Proc. Nat. Acad. Sci. U.S.A. **9** (1923), 90–93.

[Le] I. W. Lewis, *Continuum theory problems*, Topology Proc. **8** (1983), 361–394.

[M] R. Manka, *Association and fixed points*, Fund. Math. **91** (1976), 105–121.

[Ma] Sister J. L. Mark, *The fixed-point question for bounded nonseparating plane continua*, Doctor of Education Dissertation, Oklahoma State Univ., Stillwater, Oklahoma, 1968.

[Mh1] M. M. Marsh, *A fixed point theorem for inverse limits of fans*, Proc. Amer. Math. Soc. **91** (1984), 139–142.

[Mh2] ____, *A fixed point theorem for inverse limits of simple n-ods*, Topology Appl. **24** (1986), 213–216.

[Mi1] P. Minc, *An extension of the Lefschetz fixed point theorem to some plane continua*, Bull. Acad. Polon. Sci. **20** (1972), 871–878.

[Mi2] ____, *Generalized retracts and the Lefschetz fixed point theorem*, Bull. Acad. Polon. Sci. **25** (1977), 291–299.

[Mi3] ____, *A fixed point theorem for weakly chainable plane continua*, Trans. Amer. Math. Soc. (to appear).

[Mo] L. Mohler, *The fixed-point property for homeomorphisms of 1-arcwise connected continua*, Proc. Amer. Math. Soc. **52** (1975), 451–456.

[MO] L. Mohler and L. G. Oversteegen, *Open and monotone fixed point free maps on uniquely arcwise connected continua*, Proc. Amer. Math. Soc. **95** (1986), 476–482.

[Mr] R. L. Moore, *Concerning triodic continua in the plane*, Fund. Math. **13** (1929), 261–263.

[OR1] L. G. Oversteegen and J. T. Rogers, Jr., *An inverse limit description of an atriodic tree-like continuum and an induced map without a fixed point*, Houston J. Math. **6** (1980), 549–564.

[OR2] ____, *Fixed-point-free maps on tree-like continua*, Topology Appl. **13** (1982), 85–95.

[S] K. Sieklucki, *On a class of plane acyclic continua with the fixed point property*, Fund. Math. **63** (1968), 257–278.

[St] J. Stallings, *Fixed point theorems for connectivity maps*, Fund. Math. **47** (1959), 249–263.

[Y] G. S. Young, *Fixed-point theorems for arcwise connected continua*, Proc. Amer. Math. Soc. **11** (1960), 880–884.

DEPARTMENT OF MATHEMATICS, CALIFORNIA STATE UNIVERSITY, SACRAMENTO, CA 95819

Contemporary Mathematics
Volume 117, 1991

A Technique for Constructing Examples

JUDY KENNEDY

Exotic invariant continua are apparently common occurrences in discrete dynamical systems. This leads quite naturally to questions as to exactly which continua come about this way, and to what kinds of dynamical behavior the continua will support. It is the second of these questions that will be considered here.

Probably the simplest, most natural approach to constructing a homeomorphism is the following: Partition the spaces that are involved into finite collections of sets, and then assign the sets in the first space to the sets in the second space in the manner desired. These partitions and set assignments give the first approximation to the homeomorphism. To obtain an improved approximation, partition the spaces again, obtaining refinements of the first partitions, and then assign sets to sets again, keeping these assignments consistent with those already made in the first step. Continue this process, getting at the n level partitions of the space that refine the partition chosen at the $n-1$ level, and then assign sets to sets once more, being consistent with the choices already made. The homeomorphism approximation improves with each new level, and the end result (if all this is done with a fair amount of care) is a homeomorphism from the first space to the second which has the properties desired. This idea has been used by many topologists to construct homeomorphisms. See, for example, [**Bi1**], [**Bi2**], [**L1**], [**L2**], and [**KR**].

For the past several years I have been working on problems that come largely from Marcy Barge. He wanted to know what kinds of dynamical behavior different indecomposable continua, particularly pseudoarcs, would admit. I have used various versions of this homeomorphism construction technique to answer several of his questions, and have obtained some other examples along the way. These continua can admit certain types of complicated dynamical behavior. (See [**K1**], [**K2**], [**K3**], [**K5**], and [**K6**].) I have also recently used the technique to give general topological proofs to some

---

1980 *Mathematics Subject Classification* (1985 *Revision*). Primary 54F20; Secondary 54H20.
The final version of this paper will be submitted for publication elsewhere.

of the basic aspects of Carathéodory's theory of prime ends. In fact, one of Carathéodory's main theorem was generalized slightly in this way, and both a Schoenflies theorem and a theorem of Rutt were obtained as corollaries. (See [K4].) Carathéodory's theory has received a fair amount of attention from workers in dynamics lately because it has been very effectively applied to some dynamical problems in the plane. (See [AY], [BG], [M1], [M2], and [W].)

There are other methods for producing examples of dynamical systems, of course. A notable one of these is the inverse limit approach. This has been used very effectively, and sometimes quite elegantly, by R. Williams [W1, W2, W3], M. Barge and J. Martin [BM1, BM2, BM3, BM4], and P. Minc and W. R. R. Transue [MT]. The inverse limit approach does have at least one disadvantage: not every dynamical system or homeomorphism can be constructed in this way. Perhaps the homeomorphism-construction-by-approximation approach is rather more intuitive, too. However, both approaches share one common feature: both can be used to yield a homeomorphism and a continuum simultaneously. Rather than taking a space and then imposing a map onto it, the system comes into being as a whole, as a little machine ready to be switched on. I find something pleasing about this.

My plan, then, is this: The first section contains a brief history of indecomposable continua in dynamical systems as well as some background information; the second section gives a survey of the examples obtained thus far. In the second section, some of the details of the construction of a chaotic homeomorphism on the simplest indecomposable continuum will be given to illustrate an actual application of the method. (The resulting system is most probably not a new one.) Although the constructions and proofs will not be given for the more involved examples and results, references will be given.

I. Background

A continuum is a compact, connected, metric space. An indecomposable continuum is one with the property that each of its proper subcontinua is nowhere dense in the continuum. G. D. Birkhoff [B] is responsible for what is perhaps the earliest example of such a continuum in a dynamical system. In 1932, he constructed a homeomorphism on an annulus with an unusual invariant subcontinuum C. This continuum C has the property that it is the boundary set for both an open set U in the annulus containing the inner boundary circle, and an open set V containing the outer boundary circle. The continuum C is unusual in that the rotation number of a dense set of points in C accessible from U, and the rotation number of a dense set of points in C accessible from V are different. However, it was not until 1934 that M. Charpentier [C] proved that the invariant set C is an indecomposable continuum. (Rotation numbers were originally defined for homeomorphisms on circles. Through the application of Carathéodory's prime ends, the concept makes sense in this context.)

A continuum K in the plane is a cofrontier if it separates the plane into exactly two open, connected sets, and no proper subcontinuum of K separates the plane. In a pair of papers appearing in 1945 and 1951, M. L. Cartwright and J. E. Littlewood studied the forced van der Pol equation. (See [CL1] and [CL2].) At certain parameter values for that equation, there is an associated orientation-preserving homeomorphism of the plane possessing stable periodic orbits of periods $2n - 1$ and $2n + 1$ (for some positive integer n). The periodic points lie in an invariant cofrontier that contains no fixed points. Cartwright and Littlewood noticed the similarities between their invariant continuum and the one Birkhoff had obtained, and they conjectured that this cofrontier contained an indecomposable continuum.

In 1988, Marcy Barge and Richard Gillette [BG] obtained two theorems giving conditions under which indecomposable continua occur as invariant cofrontiers for orientation-preserving homeomorphisms on the plane. A corollary to their Theorem B gives the result that this invariant continuum of Cartwright and Littlewood is an indecomposable continuum, thereby "giving substance to the correspondence between complicated dynamical properties of plane homeomorphisms and exotic topology in invariant plane separating continua." Perhaps the most famous example of an indecomposable continuum occurring as an invariant continuum in a dynamical system is the stable (or unstable) manifold for Smale's horseshoe map [S]. This continuum is known to topologists as K_2, the Knaster bucket handle. The Knaster bucket handle is a nonseparating plane continuum with dense arc-components.

Marcy Barge and Joe Martin [BM1] prove in a paper entitled "Chaos, Periodicity, and Snakelike Continua," that if $f: I \to I$ is a continuous surjection on the interval which admits a point of period three, then the associated inverse limit space (I, f) contains an indecomposable continuum. Thus, paraphrasing J. Yorke and T.-Y. Li's "Period Three Implies Chaos" [LY], M. Barge and J. Martin have shown that period three implies indecomposable continua, too. (For more information on inverse limits, see J. Martin's article in this volume.) Let me mention one final example: M. Handel [H] has shown there is an area-preserving C^∞ diffeomorphism on the plane which admits an invariant continuum known as a pseudocircle as a minimal set. If he deletes his area-preserving requirement, then Handel obtains the pseudocircle as an attractor and minimal set for a C^∞ diffeomorphism. The pseudocircle is a plane separating indecomposable continuum which contains no arcs. (See [R], [F], and [KR] for more information about this continuum.)

If $F: X \to X$ is a homeomorphism, and A is a closed subset of X such that $F(A) = A$, then A is an attractor for F if there is some open set u such that $u \supseteq \overline{F(u)}$ and $\bigcap_{n=1}^\infty F^n(u) = A$. Suppose that T is a homeomorphism from X to itself. Then

(1) T is transitive if for each pair u, v of open sets of X, there is some integer n such that $T^n(u) \cap v \neq \varnothing$;

(2) T is minimal if for each $x \in X$, $O_T(x) = \{T^n(x) | n \in \mathbb{Z}\}$ is dense in X;

(3) T has sensitive dependence on initial conditions if there is some $\delta > 0$ such that if $x \in X$ and u is an open set that contains x, then there are some point y in u and positive integer n such that $d(T^n(x), T^n(y)) > \delta$;

(4) if $p \in X$ such that for some positive integer n, $T^n(p) = p$, then p is a periodic point for T;

(5) if T is transitive, has sensitive dependence on initial conditions, and has a dense set of periodic points, then T is said to be chaotic;

(6) a point $p \in X$ is said to be wandering under T if there is an open set o containing p such that $\{T^n(o)|n \in \mathbb{Z}\}$ is a collection of mutually disjoint open sets;

(7) $\Omega(T) = \{p \in X|p$ is not wandering under $T\}$; and

(8) a point p is an attracting (repelling) point for T if there is an open set o containing p such that if $x \in o$, then $\lim_{i \to \infty} T^i(x) = p(\lim_{i \to \infty} T^{-i}(x) = p)$.

A chain $C = \{c_0, c_1, \ldots, c_n\}$ is a finite collection of sets with the property that $\overline{c}_i \cap \overline{c}_j \neq \varnothing$ iff $|i - j| \leq 1$. A circular chain $C = \{c_0, c_1, \ldots, c_n\}$ is a finite collection of sets with the property that $\overline{c}_i \cap \overline{c}_j \neq \varnothing$ iff $|i - j| \leq 1$ or $|i - j| = n$. A continuum X is said to be chainable, arclike, or snakelike if for each $\varepsilon > 0$, there is a chain of sets of diameter less than ε that covers X. A continuum X is circularly chainable if for each $\varepsilon > 0$, there is a circular chain of sets of diameter less than ε that covers X. The members of a chain or a circular chain are called links.

A continuum is hereditarily indecomposable if each of its subcontinua is indecomposable. A pseudoarc may be characterized as a hereditarily indecomposable, chainable continuum. All chainable continua are nonseparating plane continua that have the fixed point property [Ha]. The Knaster continua are also indecomposable chainable continua, but since these continua have dense arc-components, they are not hereditarily indecomposable. (See [B1], [B2], and [Wa] for more details.) The pseudocircle is a circularly chainable, hereditarily indecomposable continuum that is also contained in the plane. (All circularly chainable continua in the plane separate the plane, but not all circularly chainable continua may be embedded in the plane. The solenoids are indecomposable, circularly chainable, nonplanar continua (except the circle itself).) Note that a hereditarily indecomposable continuum contains no arcs. Most continua in the plane are pseudoarcs: more precisely, the pseudoarcs form a dense G_δ-subset of the space of all continua in the plane (in the topology generated by the Hausdorff metric). (See [Bi1], [Bi2], [Bi3], [Bi4], [L1], [L2], [K5], [K6], [Kr], [KM], and [OT] for more information. Also, see the article by Wayne Lewis in this volume.) There is a homeomorphism on the plane for which the pseudoarc is an attractor, and on which the homeomorphism is chaotic. (This follows from results in [BM2] and [MT].)

In the sequel, P denotes the pseudoarc, $I = [0, 1]$, $\mathbb{Z}$ is the integers, $\mathbb{N}$ is the positive integers, and $\hat{\mathbb{N}} = \mathbb{N} \cup \{0\}$. All spaces are compact, metric spaces, and if X is a compact, metric space, then $H(X)$ denotes its group of self-homeomorphisms. If A is a collection of sets, then A^* denotes the union of the sets in A. If B is a set in the space X, then ∂B denotes the boundary of B in X and B° denotes the interior of B in X. If $A = \overline{A} \subseteq X$, $B \subseteq A$, then $\partial_A B$ denotes the boundary of B in the subspace A. If A and B are collections of sets, then $A \vee B = \{a \cap b \mid a \in A, \ b \in B, \ \text{and} \ a^\circ \cap b^\circ \neq \varnothing\}$. If $m, n \in \mathbb{Z}$, $m \leq n$, then $I[m, n] = \{m, m+1, \ldots, n\}$. $I[m, n]$ will be called an integer interval.

If X is a compact metric space and $C = \{c(0), c(1), \ldots, c(m)\}$ is a finite collection of sets of X which consists of closed, regular neighborhoods, then C is a tiling iff $c(i) \cap c(j) = \partial c(i) \cap \partial c(j)$ for each $i \neq j$ in $\{0, 1, \ldots, m\}$. (A closed neighborhood B is regular if $\overline{(B^\circ)} = B$. An open neighborhood B is regular if $(\overline{B})^\circ = B$.) If $f \colon \{0, 1, \ldots, m\} \twoheadrightarrow \{0, 1, \ldots, n\}$ is a surjection, then f is a weak pattern on $\{0, 1, \ldots, n\}$. If f has the additional property that $|f(i) - f(i+1)| \leq 1$ for $i \in I[0, m-1]$, then f is a pattern. If $G = \{g(0), g(1), \ldots, g(m)\}$ and $H = \{h(0), h(1), \ldots, h(n)\}$ are tilings of the compact metric space X, then the statement that G follows the (weak) pattern f in H means that $g(i) \subseteq h(f(i))$ for each $i \in \{0, 1, \ldots, m\}$.

The following is one version of the theorem that lies at the heart of the construction technique being discussed.

THEOREM 1. *Suppose that X and Y are compact metric spaces and G_1, $G_2, \ldots$ and $H_1, H_2, \ldots$ are sequences of tilings of X and Y, respectively, such that*

(1) *for each i, $G_i = \{g(i, 0), g(i, 1), \ldots, g(i, a(i))\}$ and $H_i = \{h(i, 0), h(i, 1), \ldots, h(i, a(i))\}$;*
(2) *if $A \subseteq \{0, 1, \ldots, a(i)\}$, then $\bigcap_{j \in A} g(i, j) \neq \varnothing$ iff $\bigcap_{j \in A} h(i, j) \neq \varnothing$;*
(3) *$\lim_i \operatorname{mesh} G_i = \lim_i \operatorname{mesh} H_i = 0$; and*
(4) *for each i, G_{i+1} follows the weak pattern η_i in G_i, and H_{i+1} follows η_i in H_i.*

Then for $x \in X$, there is an infinite sequence $j(x, 1), j(x, 2), \ldots$ of integers such that

(1) *$x \in g(i, j(x, i))$, and*
(2) *$\eta_i(j(x, i+1)) = j(x, i)$.*

Further, if we define $T(x) = \bigcap_{i < \infty} h(i, j(x, i))$ for $x \in X$, then $T \colon X \to Y$ is a homeomorphism [K1].

The examples that follow utilize a modified version of the preceding theorem. That modified version is stated below.

THEOREM 2. *Suppose X is a compact, metric space, and $X_1, X_2, \ldots$ is a nested sequence of closed subsets of X. If $G_1, G_2, \ldots$ and $H_1, H_2, \ldots$ are sequences of tilings in X such that*

(1) *for each i, $G_i = \{g(i, 0), g(i, 1), \ldots, g(i, a(i))\}$ and $H_i = \{h(i, 0), h(i, 1), \ldots, h(i, a(i))\}$;*
(2) *G_i covers X_i and H_i covers X_{i+1};*
(3) *if $A \subseteq I[0, a(i)]$, then $\bigcap_{j \in A} g(i, j) \neq \varnothing$ iff $\bigcap_{j \in A} h(i, j) \neq \varnothing$;*
(4) *$\lim_i \operatorname{mesh} G_i = \lim_i \operatorname{mesh} H_i = 0$; and*
(5) *for each i, G_{i+1} follows the weak pattern η_i in G_i, and H_{i+1} follows the weak pattern η_i in H_i.*

Then for $x \in \bigcap X_i = \hat{X}$, there is an infinite sequence $j(x, 1), j(x, 2), \ldots$ of integers such that

(1) *$x \in g(i, j(x, i))$, and*
(2) *$\eta_i(j(x, i + 1)) = j(x, i)$.*

Further, if we define $T(x) = \bigcap_{i < \infty} h(i, j(x, i))$ for $x \in X$, then $T: \hat{X} \to \hat{X}$ is a homeomorphism.

II. Examples and results

THEOREM 3. *Suppose X is a compact metric space that satisfies the hypotheses of Theorem 2. (The notation used below has the meaning assigned in that theorem.) In addition, suppose that*

(6) *for each odd i, H_{i+1} follows the weak pattern ξ_i in G_i;*
(7) *for each even i, G_{i+1} follows the weak pattern ξ_i in H_i;*
(8) *for each odd i, and for each $k, j \in I[0, a(i + 1)]$, $\xi_i(k) = \eta_i(j)$ iff $h(i, k)^\circ \cap g(i, j)^\circ \neq \varnothing$; and*
(9) *for each even i, and for each $k, j \in I[0, a(i + 1)]$, $\xi_i(k) = \eta_i(j)$ iff $g(i, k)^\circ \cap h(i, j)^\circ \neq \varnothing$.*

Then for each $x \in \hat{X}$, there is an infinite sequence $j(x, 1), j(x, 2), \ldots$ such that for each i

(1) *$x \in g(i, j(x, i))$, and*
(2) *$\eta_i(j(x, i + 1)) = j(x, i)$.*

Further, if for $x \in \hat{X}$, $T(x) = \bigcap_{i=1}^\infty h(i, j(x, i))$, then $T: \hat{X} \to \hat{X}$ is a chaotic homeomorphism on $\hat{X}$. (See [K1] for the proof.)

EXAMPLE 4. A chaotic homeomorphism on K_2. (This is most probably just another way to describe the shift on K_2. See Figures 1 and 2 for help in understanding the construction.)

Begin by choosing $G_1 = \{g(1, 0), g(1, 1)\}$ to be a chain and tiling in the plane, and choose $H_1 = \{h(1, 0), h(1, 1)\}$ to be a chain and tiling such that

(1) $H_1^* \subseteq G_1^{*\circ}$,
(2) for i and j in $I[0, 1]$, $g(1, i) \cap h(1, j)$ is homeomorphic to the unit disk D, and
(3) $\partial h(1, 0) \cap \partial h(1, 1) \subseteq g(1, 1)^\circ$.

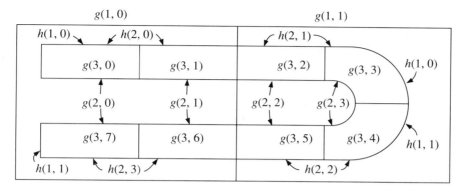

FIGURE 1

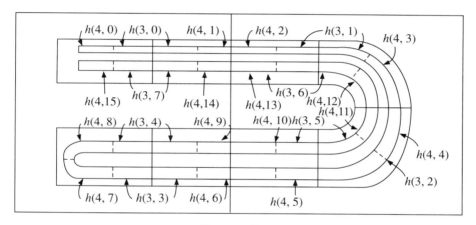

FIGURE 2

Let $H_2 = G_1 \vee H_1$. Then

$$H_2 = H_2[0, 3] = \{h(2, 0), h(2, 1), h(2, 2), h(2, 3)\}$$

is a chain and tiling that follows the pattern ξ_1 in H_1 and the pattern η_1 in G_1, where we define for $i \in I[0, 3]$,

$$\xi_1(i) = j \quad \text{iff } i \in \{2j, 2j+1\},$$

$$\eta_1(i) = \begin{cases} i \bmod 2 & \text{if } [[i/2]] \text{ is even}, \\ 1 - i \bmod 2 & \text{if } [[i/2]] \text{ is odd}. \end{cases}$$

Choose the chain and tiling $G_2 = G_2[0, 3] = \{g(2, 0), g(2, 1), g(2, 2), g(2, 3)\}$ as follows:

(4) G_2 follows ξ_1 in G_1.
(5) The sets in $G_2 \vee H_2$ are homeomorphic to D.
(6) $G_2^* = H_2^*$.
(7) If $i, j \in I[0, 3]$ and $\eta_1(j) = \xi_1(i)$, then $g(2, j)° \cap h(2, i)° \neq \varnothing$.

If $G_3 = G_2 \vee H_2$, $G_3 = G_3[0, 2^3 - 1] = \{g(3, 0), \ldots, g(3, 7)\}$ is a chain and, with the appropriate listing of links of G_3, G_3 follows ξ_2 in G_2 and η_2 in H_2, where we define for $i \in I[0, 7]$,

$$\eta_2(i) = j \quad \text{iff } i \in I[2j, 2j+1],$$

$$\xi_2(i) = \begin{cases} i \bmod 4 & \text{if } [[i/4]] \text{ is even}, \\ 3 - i \bmod 4 & \text{if } [[i/4]] \text{ is odd}. \end{cases}$$

(Note that "ξ" and "η" have switched roles in a sense.) Now choose the chain and tiling $H_3 = H_3[0, 7] = \{h(3, 0), \ldots, h(3, 7)\}$ as follows:

(8) H_3 follows ξ_2 in H_2.

(9) The sets in $H_3 \vee G_3$ are homeomorphic to D.

(10) $H_3^* \subseteq G_3^{*\circ}$.

(11) For i and $j \in I[0, 7]$, if $\eta_2(i) = \xi_2(j)$, then $h(3, j)^\circ \cap g(3, i)^\circ \neq \varnothing$.

Continue the inductive process that has been begun, obtaining sequences $G_1, G_2, \ldots$ and $H_1, H_2, \ldots$ of chains and tilings which (assuming the new links are chosen with care at each level) have the following properties:

(12) $\lim \operatorname{mesh} G_i = \lim \operatorname{mesh} H_i = 0$.

(13) $G_1^* \supseteq G_1^{*\circ} \supseteq H_1^* = H_2^* = G_2^* = G_3^* \supseteq G_3^{*\circ} \supseteq H_3^* = \cdots$.

(14) For $i \in \mathbb{N}$, H_{i+1} follows ξ_i in H_i, and G_{i+1} follows ξ_i in G_i.

(15) For i even, G_{i+1} follows η_i in H_i, and for i odd, H_{i+1} follows η_i in G_i.

(16) For k odd, $i \in I[0, 2^{k+1} - 1]$,

$$\xi_k(i) = j \quad \text{iff } i \in \{2j, 2j+1\},$$

$$\eta_k(i) = \begin{cases} i \bmod 2^k & \text{if } [[i/2^k]] \text{ is even}, \\ (2^k - 1) - i \bmod 2^k & \text{if } [[i/2^k]] \text{ is odd}. \end{cases}$$

(17) For k even, $i \in I[0, 2^{k+1} - 1]$,

$$\eta_k(i) = j \quad \text{iff } i \in \{2j, 2j+1\},$$

$$\xi_k(i) = \begin{cases} i \bmod 2^k & \text{if } [[i/2^k]] \text{ is even}, \\ (2^k - 1) - i \bmod 2^k & \text{if } [[i/2^k]] \text{ is odd}. \end{cases}$$

(18) If $i, j \in I[0, 2^{k+1} - 1]$ and $\eta_k(i) = \xi_k(j)$, then if k is even,

$$g(k+1, i)^\circ \cap h(k+1, j)^\circ \neq \varnothing,$$

and if k is odd,

$$g(k+1, j)^\circ \cap h(k+1, i)^\circ \neq \varnothing.$$

Then $\bigcap H_i^* = \bigcap G_i^*$, and this intersection is a nondegenerate, chainable, indecomposable continuum, which is K_2. Also, a chaotic homeomorphism $T: K_2 \to K_2$ has been induced (Theorem 3). $\square$

THEOREM 4. *If P' is a subcontinuum of the pseudoarc P, h' is a homeomorphism from P' onto itself, and o is an open set in P that contains P', then there is a homeomorphism h from P onto itself such that $h|P' = h'$ and $h(x) = x$ for $x \notin o$. (See* [K5].)

EXAMPLE 5. Suppose that X is a continuum with the property that each of its proper subcontinua is a pseudoarc. In addition, suppose that (1) $n_1, n_2, \ldots$ is a sequence of positive integers, (2) $o_1, o_2, \ldots$ is a sequence of disjoint open sets in X with $\lim_i \operatorname{diam} o_i = 0$, and (3) for $i \in \mathbb{N}$, P is a nondegenerate subcontinuum contained in o_i with $x_i \in P_i$. Then there is a homeomorphism $h \in H(X)$ such that for each $i \in \mathbb{N}$, $h(P_i) = P_i$, x_i has period n_i under h, and $h(x) = x$ for $x \notin \bigcup o_i$. (See [K5].)

EXAMPLE 6. There is a transitive homeomorphism T on the pseudoarc P with the property that T factors over the tent map on the interval. (The tent map on the interval is transitive on the interval. Transitive in this context means that if t denotes the tent map, then there is some x in I such that $\{t^n(x)|n \in \mathbb{N}\}$ is dense in I.) (See [K6].)

EXAMPLE 7. There is a homeomorphism T on the pseudoarc P with the properties that T factors over the tent map on the interval and T admits wandering points. (See [K6].)

EXAMPLE 8. There is a homeomorphism h on P such that (1) h admits exactly one attracting point p and one repelling point q; (2) $\Omega(h) = \{p, q\}$ with p and q fixed points for h; and (3) h factors over the map $f \in H(I)$ defined by $f(x) = x^2$. (See [K2].)

EXAMPLE 9. There is a homeomorphism h on P such that $\Omega(h) = \{p\}$ for some point p in P. (Note that it follows that $h(p) = p$, and for $x \in P\setminus\{p\}$, $\lim_{i \to \infty} h^i(x) = p = \lim_{i \to \infty} h^{-i}(x)$.) (See [K2].)

THEOREM 10. *If C is a Cantor set and h is a homeomorphism from C onto itself, then C can be embedded in a pseudoarc P so that the homeomorphism on the image of C in P induced by h can be extended to a homeomorphism f on P. (See* [K3].)

COROLLARY 11. *The pseudoarc admits a homeomorphism with positive entropy. (See* [K3].)

EXAMPLE 12. The pseudoarc admits a chaotic homeomorphism. (See [K1]. This has subsequently been proven, using other methods, by P. Minc and W. R. R. Transue [MT].)

REFERENCES

[AY] K. T. Alligood and J. A. Yorke, *Accessible saddles on fractal basin boundaries*, Preprint.

[B1] M. Barge, *Horseshoe maps and inverse limits*, Pacific J. Math. **121** (1986), 29–39.

[B2] ———, *The topological entropy of homeomorphisms of Knaster continua*, Houston J. Math. **13** (1987), 465–485.

[BG] M. Barge and R. Gillette, *Indecomposability and dynamics of invariant plane separating continua*, Preprint.

[**BM1**] M. Barge and J. Martin, *Chaos, periodicity, and snakelike continua*, Trans. Amer. Math. Soc. **289** (1985), 355–365.

[**BM2**] ____, *The construction of global attractors*, Preprint.

[**BM3**] ____, *Dense orbits on the interval*, Michigan Math. J. **34** (1987), 3–11.

[**BM4**] ____, *Dense periodicity on the interval*, Proc. Amer. Math. Soc. **94** (1985), 731–735.

[**Bi1**] R. H. Bing, *Concerning hereditarily indecomposable continua*, Pacific J. Math. **1** (1951), 43–51.

[**Bi2**] ____, *A homogeneous indecomposable plane continuum*, Duke Math. J. **15** (1948), 729–742.

[**Bi3**] ____, *Each homogeneous nondegenerate chainable continuum is a pseudoarc*, Proc. Amer. Math. Soc. **10** (1959), 345–346.

[**Bi4**] ____, *Snakelike continua*, Duke Math. J. **18** (1951), 653–663.

[**B**] G. D. Birkhoff, *Sur quelques courbes fermees remarquables*, Bull. Soc. Math. France **60** (1932), 1–26.

[**CL1**] M. L. Cartwright and J. E. Littlewood, *On non-linear differential equations of the second order. I. The equation* $\ddot{y} = k(1 - y^2)\dot{y} + y = b\lambda k \cos(\lambda t + \alpha)$, k *large*, J. London Math. Soc. **20** (1945), 180–189.

[**CL2**] ____, *Some fixed point theorems*, Ann. Math. **54** (1951), 1–37,

[**C**] M. Charpentier, *Sur quelques propriétés des courbes de M. Birkhoff*, Bull. Soc. Math. France **62** (1934), 193–224.

[**F**] L. Fearnley, *The pseudocircle is not homogeneous*, Bull. Amer. Math. Soc. **75** (1969), 554–558.

[**Ha**] O. H. Hamilton, *A fixed point theorem for pseudoarcs and certain other metric continua*, Proc. Amer. Math. Soc. **2** (1951), 173–174.

[**H**] Michael Handel, *A pathological area preserving C^∞ diffeomorphism of the plane*, Proc. Amer. Math. Soc. **86** (1982), 163–168.

[**K1**] J. Kennedy, *The construction of chaotic homeomorphisms on chainable continua*, Topology Appl. (to appear).

[**K2**] ____, *Examples of homeomorphisms on pseudoarcs that admit wandering points*, Topology Appl. (to appear).

[**K3**] ____, *Positive entropy homeomorphisms on the pseudoarc*, Michigan Math. J. **36** (1989), 181–191.

[**K4**] ____, *A really topological treatment of some aspects of Carathéodory's theory of prime ends*, preprint.

[**K5**] ____, *Stable extensions of homeomorphisms on the pseudoarc*, Trans. Amer. Math. Soc. **310** (1988), 167–178.

[**K6**] ____, *A transitive homeomorphism on the pseudoarc which is semiconjugate to the tent map*, Trans. Amer. Math. Soc. (to appear).

[**KR**] J. Kennedy and J. T. Rogers, Jr., *Orbits of the pseudocircle*, Trans. Amer. Math. Soc. **296** (1986), 327–340.

[**Kr**] J. Krasinkiewicz, *Mapping properties of hereditarily indecomposable continua*, Preprint.

[**KM**] J. Krasinkiewicz and P. Minc, *Mappings onto indecomposable continua*, Bull. Acad. Pol. Sci. **25** (1977), 675–680.

[**L1**] W. Lewis, *Most maps of the pseudoarc are homeomorphisms*, Proc. Amer. Math. Soc. **91** (1984), 147–153.

[**L2**] ____, *Stable homeomorphisms of the pseudoarc*, Canad. J. Math. **31** (1979), 363–374.

[**LY**] T.-Y. Li and J. A. Yorke, *Period three implies chaos*, Amer. Math. Monthly **82** (1975), 985–992.

[**M1**] J. N. Mather, *Area preserving twist homeomorphisms of the annulus*, Comment. Math. Helv. **54** (1979), 397–404.

[**M2**] ____, *Invariant subsets for area preserving homeomorphisms of the annulus*, Adv. Math. Suppl. Stud., vol. 7B.

[**MT**] P. Minc and W. R. R. Transue, *A transitive map on* [0 , 1] *whose inverse limit is the pseudoarc*, preprint.

[OT] L. G. Oversteegen and E. D. Tymchatyn, *On hereditarily indecomposable continua*, Geometric and Algebraic Topology, Banach Centre Publ., vol. 18, PWN, Warsaw, 1986, pp. 403–413.

[R] J. T. Rogers, Jr., *The pseudocircle is not homogeneous*, Trans. Amer. Math. Soc. **148** (1970), 417–428.

[S] S. Smale, *Differentiable dynamical systems*, Bull. Amer. Math. Soc. **73** (1967), 747–817.

[W] R. Walker, *Basin boundaries with irrational rotation on prime ends*, Trans. Amer. Math. Soc. (to appear).

[Wa] W. T. Watkins, *Homeomorphic classification of certain inverse limit spaces with open bonding maps*, Pacific J. Math. **103** (1982), 589–601.

[Wi1] R. F. Williams, *Expanding attractors*, Inst. Hautes Études Sci. Publ. Math. **43** (1974), 169–203.

[Wi2] ____, *One-dimensional nonwandering sets*, Topology **6** (1967), 473–487.

[Wi3] ____, *The structure of Lorenz attractors*, Inst. Hautes Études Sci. Publ. Math. **50** (1979), 101–152.

DEPARTMENT OF MATHEMATICAL SCIENCES, UNIVERSITY OF DELAWARE, NEWARK, DELAWARE 19716

Contemporary Mathematics
Volume 117, 1991

Continuum Theory and Dynamics Problems

WAYNE LEWIS

Most of the following problems concern the pseudo-arc. Several have been posed previously, either in the form posed here or in variations. No effort has been made here to fully trace the history of each problem or determine who first posed it. I will be glad to assist anyone interested in a specific problem in determining its history. Only brief background is provided here on these problems. I am also willing to assist in determining this more fully.

Q1. (M. Barge) Which homeomorphisms of the pseudo-arc can occur as restrictions of diffeomorphisms of $\mathbb{R}^2$ to an invariant pseudo-arc?

Q2. (M. Barge) If $f: \mathbb{R}^2 \to \mathbb{R}^2$ is a homeomorphism with an invariant pseudo-arc P, is $f|_P$ conjugate to an ε-homeomorphism $g|_P$ for some $g: \mathbb{R}^2 \to \mathbb{R}^2$?

Lewis has shown that every homeomorphism of the pseudo-arc is conjugate to an ε-homeomorphism. It seems improbable that such conjugacy can be achieved while preserving essential extendability to $\mathbb{R}^2$.

Q3. (J. Walsh, M. Bestvina) Does there exist an embedding of the pseudo-arcs in $\mathbb{R}^2$ such that under the action of the group of extendable homeomorphisms each (accessible) point has a dense orbit?

Q4. What are necessary and sufficient conditions for a map between closed subsets of the pseudo-arc to be extendable to a map of the entire pseudo-arc onto itself?

Lewis has shown that if the closed subsets are proper and have only a finite number of components, then this is always possible. He has also determined necessary and sufficient conditions for a homeomorphism between closed subsets of the pseudo-arc to extend to a homeomorphism of the pseudo-arc onto itself.

Q5. (E. Burgess) Does there exist an embedding of the pseudo-arc in $\mathbb{R}^2$ with no nonidentity extendable homeomorphisms?

1980 *Mathematics Subject Classification* (1985 *Revision*). Primary 54F15, 54F20, 54H20.
This paper is in final form and no version of it will be submitted for publication elsewhere.

Q6. (B. Brechner, W. Lewis) Do there exist stable homeomorphisms of the pseudo-arc which are essentially extendable to $\mathbb{R}^2$? How many up to conjugacy?

Lewis has shown the existence of stable homeomorphisms of the pseudo-arc.

Q7. (B. Brechner) Let M be a particular embedding of the pseudo-arc in $\mathbb{R}^2$ and let G be the group of extendable homeomorphisms of M. Does G characterize the embedding?

Q8. (L. Ward) Does the pseudo-arc P have the complete invariance property, i.e., is every nonempty closed subset the fixed point set of some self-map of P? continuous surjection of P? homeomorphism of P?

Q9. (W. Kuperberg) A continuum X is *pseudo-contractible* if there exists a continuum Y and a map $f: X \times Y \to X$ such that for some $a, b \in Y$ and $x_0 \in X$, $f(x, a) = x$ and $f(x, b) = x_0$ for all $x \in X$. Is it true that the pseudo-arc is not pseudo-contractible?

Q10. Is the collection of homeomorphisms of the pseudo-arc P essentially extendable to homeomorphisms of $\mathbb{R}^2$ of first category in $H(P)$, the space of all homeomorphisms of P?

Q11. Is the collection of homeomorphisms of the pseudo-arc P, which are essentially extendable to homeomorphisms of $\mathbb{R}^2$, dense in $H(P)$?

Q12. Does every embedding of the pseudo-arc P in $\mathbb{R}^2$ have only a nowhere dense collection (in $H(P)$) of extendable homeomorphisms?

Q13. Does there exist a pseudo-arc in the plane which has a subcontinuum of each possible embedding type?

Q14. Do any two pseudo-arcs in the plane have nondegenerate subcontinua which are equivalently embedded?

Q15. Are most pseudo-arcs in the plane of one embedding type?

Q16. (E. Burgess) Are there $c = 2^{\omega_0}$ distinct embeddings of the pseudo-circle in $\mathbb{R}^2$? of every nondegenerate hereditarily indecomposable plane continuum?

Lewis and Smith have, independently, shown that there are $c = 2^{\omega_0}$ distinct embeddings of the pseudo-arc in $\mathbb{R}^2$, including, for each positive integer n, one with exactly n accessible composants and one with each accessible point in a different composant.

Q17. (M. Barge) Does the pseudo-arc, for each infinite set of integers including one, admit a homeomorphism with points of exactly these periods and no others?

Q18. (M. Barge) Does every homeomorphism of the pseudo-arc (or any chainable continuum) of positive entropy have periodic points of infinitely many different periods? Equivalently, must a positive entropy homeomorphism of a chainable continuum have a nonfixed periodic point?

Q19. (M. Barge) Is every positive real number the entropy of some homeomorphism of the pseudo-arc?

Q20. (M. Barge) If the homeomorphism $h: P \to P$ of the pseudo-arc P is transitive, is the entropy of h positive?

Q21. (E. Burgess) If C is a nondegenerate chainable continuum with the property that all embeddings of C in $\mathbb{R}^2$ are equivalent, must C be an arc?

Q22. Which self-maps of a triod, or tree, are semiconjugate to homeomorphisms of the pseudo-arc?

Lewis has shown that every self-map of a chainable continuum is semiconjugate to a homeomorphism of the pseudo-arc.

TEXAS TECH UNIVERSITY, LUBBOCK, TEXAS 79409-1042

Contemporary Mathematics
Volume **117**, 1991

The Pseudo-arc

WAYNE LEWIS

This survey is dedicated to R. H. Bing and F. Burton Jones

The pseudo-arc is the simplest nondegenerate hereditarily indecomposable continuum, though probably also the most interesting such continuum. Though it is chainable, it is also homogeneous, admits many zero-dimensional compact group actions, and has a number of interesting and useful mapping properties. It is the purpose of this article to give a brief survey of these results and of some open questions. Though we do not give proofs, these can be located in the indicated references or by combining results elsewhere. It is planned to eventually produce a more comprehensive monograph on the pseudo-arc and other hereditarily indecomposable continua with development of all essential results.

This article is divided into the following sections, with some overlap in coverage. Section 1 gives basic terminology, with Section 2 giving a description of the pseudo-arc and of its essential properties. In Sections 3 and 4 we discuss various characterizations of the pseudo-arc and its mapping properties, respectively. Section 5 presents a discussion of continuous decompositions into pseudo-arcs, while Section 6 covers homeomorphism groups. We conclude with a selection of questions in Section 7. We do not attempt completeness in this brief survey, emphasizing the basic properties of the pseudo-arc and appropriate references. We omit any extensive discussion of related continua or indication of the many places, in continuum theory and elsewhere, where the pseudo-arc has appeared as an example.

The pseudo-arc and its homeomorphisms are already known to exhibit very rich structure, though we can expect many more interesting results to be forthcoming. Indications are that this structure is much more limited when one considers embeddings of the pseudo-arc in the plane and homeomor-

1980 *Mathematics Subject Classification* (1985 *Revision*). Primary 54F20, 54F50, 54F65.

The author is supported in part by NSF grant DMS 8620338 and Texas Advanced Research Program grant 1296.

This paper is in final form and no version of it will be submitted for publication elsewhere.

phisms of the pseudo-arc which extend to homeomorphisms of the plane. That the pseudo-arc has at least some role to play in the dynamics of the plane is suggested by the use of the related continuum the pseudocircle by Handel [35].

1. Basic terminology

A *continuum* is a compact connected metric space. A continuum is *decomposable* if it can be expressed as the union of two of its proper subcontinua. Otherwise, it is *indecomposable*. A continuum is *hereditarily indecomposable* if every subcontinuum of it is indecomposable. If x is a point of the continuum X, then the *composant* of x in X is the union of all proper subcontinua of X which contain x. If x is indecomposable, then any two composants of X are disjoint or identical. The collection of all distinct composants of the nondegenerate indecomposable continuum X forms a partition of X into uncountably many subsets. Each composant of a continuum is dense, and each proper subcontinuum is contained in some composant. Thus a nondegenerate indecomposable continuum is nowhere locally connected, and an hereditarily indecomposable continuum contains no nondegenerate subcontinuum which is locally connected at any point. (Actually, such continua fail to satisfy weakenings of local connectivity such as aposyndesis. One can form a spectrum of connectivity properties ranging from hereditary local connectivity at one end to hereditary indecomposability at the other [40].) A continuum X is *chainable* if it is homeomorphic to an inverse limit of intervals, or equivalently if for each $\varepsilon > 0$ there is a map of X into an interval with each point inverse of diameter less than ε. While much work on the pseudo-arc and other chainable continua can be done in terms of inverse limits, for many purposes it is more convenient to work with chains. A *chain* is a collection $C = \{C_0, C_1, \ldots, C_n\}$ of sets such that $C_i \cap C_j \neq \varnothing$ if and only if $|i - j| \leq 1$. Most chains under consideration will be open covers. The *mesh* of a chain is the maximum diameter of its links. An *ε-chain* is a chain of mesh less than ε. A continuum X is chainable if for each $\varepsilon > 0$ there is an ε-chain of open sets covering X. Elsewhere in the literature, chainable continua have been referred to as *arc-like* or *snake-like* continua [7, 88].

Each element of a chain is called a *link*. If chain C_1 refines chain C_0, then chain C_1 is *crooked* in C_0 provided that for every p, s, i, j, where $j > i + 2$, $C_1(p) \cap C_0(i) \neq \varnothing$ and $C_1(s) \cap C_0(j) \neq \varnothing$, there exist q, r with $\mathrm{cl}(C_1(q)) \subseteq C_0(j - 1)$, $\mathrm{cl}(C_1(r)) \subseteq C_0(i + 1)$, and either $p < q < r < s$ or $p > q > r > s$. (Here $C_m(k)$ indicates the kth link of chain C_m.)

A *pattern* is a function $f: \{0, 1, \ldots, m\} \to \{0, 1, \ldots, n\}$ such that, for each $i < m$, $|f(i + 1) - f(i)| \leq 1$. Chain C_1 *follows the pattern f in chain C_0* if $f: \{0, 1, \ldots, m\} \twoheadrightarrow \{0, 1, \ldots, n\}$ where $C_1 = \{C_1(i)\}_{i \leq m}$, $C_0 = \{C_0(j)\}_{j \leq n}$, and $\mathrm{cl}(C_1(i)) \subseteq C_0(f(i))$ for each $i \leq m$. (Note that in general there will be several patterns which C_1 follows in C_0.) Chain C_0 is an *amalgamation* of chain C_1 if every link of C_0 is the union of links of C_1.

A continuum is *tree-like* if it is homeomorphic to an inverse limit of trees (i.e., acyclic finite graphs), or equivalently if for each $\varepsilon > 0$ there is a map of it onto a tree with each point inverse of diameter less than ε. Every chainable continuum is also tree-like. A continuum is *circle-like* if it is homeomorphic to an inverse limit of simple closed curves, or equivalently if for each $\varepsilon > 0$ there is a map of it onto a simple closed curve with each point inverse of diameter less than ε. Both tree-like and circle-like continua can also be defined in terms of open covers of arbitrarily small mesh whose nerves are either trees or simple closed curves. The pseudo-arc is an example of a continuum which is both chainable and circle-like.

A *curve* is a one-dimensional continuum. We will restrict our attention mainly to curves, but Bing has shown the existence of hereditarily indecomposable continua of every dimension [8].

A continuum X is *homogeneous* if for each $x, y \in X$ there is a homeomorphism h of X onto X with $h(x) = y$. X is *almost homogeneous* if for each $x \in X$ and open set U in X there exists a homeomorphism h of X onto X with $h(x) \in U$. The collection of all self-homeomorphisms of X forms a topological group, denoted $H(X)$, under composition and the sup metric. For each $x \in X$, the set $\{h(x)|h \in H\}$ is the *orbit of x under the action of $H(X)$*. If h is a particular homeomorphism of X, the set $\{h^n(x)|n \in \mathbb{Z}\}$ is the *orbit of x under h*.

A homeomorphism $h: X \to\to X$ is *stable* if there exist homeomorphisms $h_0, h_1, \ldots, h_n$ such that $h = h_n \circ h_{n-1} \circ \cdots \circ h_1 \circ h_0$, where for each $i \le n$ there exists a nonempty open set U_i such that $h_i|U_i$ is the identity. If $n = 0$, then h is *primitively stable*. (Brown and Gluck [17] called such a homeomorphism *somewhere the identity*.) If C is an open cover of the topological space X, and $x \in X$, then $\mathrm{st}(x, C)$, the *star of x in C*, is the union of the collection of elements of C which contain x.

A continuum is *weakly chainable* if it is the continuous image of a chainable continuum. Fearnley [29] and Lelek [58] have independently given a characterization of weak chainability in terms of open covers. A continuum M is *almost chainable* if for every $\varepsilon > 0$ there exists an open cover $C = \{C_0, C_1, \ldots, C_n\}$ of M such that for each $i < n$ the diameter of C_i is less than ε and for each point $x \in M$ there exists $j < n$ such that $\mathrm{dist}(x, C_j) < \varepsilon$, i.e., $\{C_0, C_1, \ldots, C_{n-1}\}$ is an ε-chain which is ε-dense in M and intersects C_n only in C_{n-1}.

A map $f: P \to Q$ between polyhedra is *ε-crooked* if for each pair of points p and q in P and arc A in P joining P to Q there exist points r and s of A such that r lies between p and s, $\mathrm{dist}(f(p), f(s)) < \varepsilon$ and $\mathrm{dist}(f(r), f(q)) < \varepsilon$.

The *span* of a continuum X is the largest ε such that there exists a subcontinuum A of $X \times X$ with $\pi_1(A) = \pi_2(A)$ and $\mathrm{dist}_X(x_1, x_2) > \varepsilon$ for each $(x_1, x_2) \in A$.

2. Description and essential properties

The first example of a nondegenerate hereditarily indecomposable continuum was described by Knaster [50] in 1922. The example he constructed was
a nested intersection of discs in the plane. In 1948 Moise [92] constructed a
nondegenerate indecomposable plane continuum with the property that it was
homeomorphic to each of its nondegenerate subcontinua. Since the arc is the
only other known nondegenerate continuum with this property, and the continuum described by Moise is also chainable, he referred to it as a *pseudo-arc*.
The property of being homeomorphic to each nondegenerate subcontinuum
has since been termed *hereditary equivalence*.

THEOREM [92]. *The pseudo-arc is a nondegenerate, indecomposable, chainable, hereditarily equivalent continuum.*

Soon afterwards, Bing [5] proved that the pseudo-arc as described by Moise
is homogeneous. Shortly thereafter Moise [93] gave his own proof of the
homogeneity of the pseudo-arc.

THEOREM [5, 93]. *The pseudo-arc is a nondegenerate, indecomposable, homogeneous, plane continuum.*

In his work on the pseudo-arc, Moise used a construction specifying, in
a fairly restrictive manner, how one chain in a defining sequence refines a
previous chain. Bing introduced a more general concept of one chain being crooked in another. He then proved a number of theorems concerning
crooked chains and amalgamations of chains. These theorems, and the techniques involved in their proofs, have been used in much subsequent work
and constructions involving the pseudo-arc. Though the pseudo-arc can also
be described in terms of inverse limits, and Mioduszewski [88, 89] has given
proofs of the basic properties and characterization of the pseudo-arc involving inverse limits and always oscillating functions, in general it has proven
more useful to consider it as given by a defining sequence of crooked chains.

THEOREM [5]. *If D, E and F are chains such that D is an amalgamation
of E and F is crooked in E, then F is crooked in D.*

THEOREM [5]. *If the chain D is crooked in the chain $E = \{E_0, E_1, \ldots, E_n\}$
and D_1 is a particular link of D, then there is a chain F such that F is
an amalgamation of D, D_i is contained in only the first link of F, each link
of F is a subset of the union of two adjacent links of E, and any link of
F containing an end link of D which intersects E_0 and E_n is a subset of
$E_0 \cup E_1$ or $E_{n-1} \cup E_n$.*

THEOREM [5]. *If $D = \{D_0, D_1, \ldots, D_n\}$ is a chain crooked in the chain
$E = \{E_0, E_1, \ldots, E_m\}$, and $D(r, s)$ is a subchain of D such that a link of*

$D(r, s)$ intersects E_m, then there is a chain F such that F is an amalgamation of D, each element of F is a subset of the union of two adjacent links of E, D_r is contained in only the first link of F, and D_s is contained only in the last.

THEOREM [6]. *If C is a chain which is an open cover of the hereditarily indecomposable continuum X, there exists a chain D refining C such that D is an open cover of X and D is crooked in C.*

From the above theorems Bing [6] was able to show that any two chainable, nondegenerate, hereditarily indecomposable continua are homeomorphic. Thus the continuum described by Moise is homeomorphic to that described earlier by Knaster.

THEOREM [6]. *Every hereditarily indecomposable chainable continuum X has a sequence of chains $\{C_i\}_{i=0}^{\infty}$ such that*
(1) *each C_i is an open cover of X,*
(2) *mesh$(C_i) < \frac{1}{2^i}$ for each i, and*
(3) *C_{i+1} is crooked in C_i for each i.*

THEOREM [6]. *Every nondegenerate, hereditarily indecomposable, chainable continuum is homeomorphic to the pseudo-arc.*

Though it is not explicitly stated in the work of Bing, the following theorem follows from the theorems proven by him.

THEOREM. *If C is a chain consisting of n links which is an open cover of the hereditarily indecomposable continuum X, and f is a pattern whose image has n elements, there exists a chain D refining C such that D is an open cover of X and D follows the pattern f in C.*

Notice that most of the above theorems do not assume that X is chainable and make no claims about the meshes of the chains involved. They apply to every hereditarily indecomposable continuum, yielding the following result.

THEOREM. *Every nondegenerate hereditarily indecomposable continuum admits a continuous surjection onto the pseudo-arc.*

In light of the above theorems, crookedness is what characterizes hereditary indecomposability. Bing generalized crooked chains by introducing the concept of an ε-crooked map between polyhedra.

THEOREM. *The continuum X is hereditarily indecomposable if and only if X is homeomorphic to the inverse limit of a sequence of polyhedra $\varprojlim (P_i, f_i^j)$ where for each positive integer i and each $\varepsilon > 0$ there exists a positive integer N such that if $j > N$ then f_i^j is an ε-crooked map from P_j to P_i.*

Bing also showed that there exists a large variety of distinct hereditarily indecomposable continua.

THEOREM [6]. *There are as many topologically different hereditarily inde-composable plane continua as there are real numbers.*

THEOREM [8]. *If M and N are two mutually exclusive continua in either $\mathbb{R}^n$, for any n, or the Hilbert cube H, there exists an hereditarily indecomposable continuum X which has exactly two complementary domains and which is irreducible with respect to separating M from N.*

THEOREM [8]. *Each $(n+1)$-dimensional continuum contains an n-dimensional hereditarily indecomposable continuum.*

The following theorems by Bing indicate when points of the pseudo-arc can be put in end links of a chain covering it.

THEOREM [5]. *If p is a point of the pseudo-arc P and $\varepsilon > 0$, there exists a chain of mesh less than ε which is an open cover of P which has its first link as the only link containing p.*

THEOREM [5]. *If p and q are points in distinct composants of the pseudo-arc P and $\varepsilon > 0$, there exists a chain of mesh less than ε which is an open cover of P and has p and q in its opposite end links.*

Using similar techniques, Lehner proved the following theorem for collections of subcontinua of the pseudo-arc.

THEOREM [57]. *Suppose $H_0, H_1, \ldots, H_n$ are nondegenerate subcontinua of the pseudo-arc P such that H_i and H_j are in different composants of P if $i \neq j$. Suppose H_i is irreducible between the points P_i and Q_i for $i \leq n$. Then for each $\varepsilon > 0$ there exists a chain D, with $\operatorname{mesh}(D) < \varepsilon$, such that D is an open cover of P, $D(P_i, Q_i)$ contains H_i for each i, $D(P_i, Q_i) \cap D(P_j, Q_j) = \varnothing$ if $i \neq j$, $D(P_1) = D_0$, $D(Q_n) = D_m$, and $D(Q_1) < D(P_2) < D(Q_2) < \cdots < D(P_n)$. (Here $D(A)$ is the first link of D containing the point A, $D(A, B)$ is the union of the links in the subchain of D from $D(A)$ through $D(B)$, D_0 is the first link of D, and D_m the last link of D.)*

The following theorem by the author gives necessary and sufficient conditions for being able to form a chain following a given pattern in another chain while including a specific point in a specific link.

THEOREM [60]. *Let $C = \{C(i)\}_{i \leq n}$ be a chain which is an open cover of the pseudo-arc P, $f: \{0, 1, \ldots, m\} \to \{0, 1, \ldots, n\}$ a pattern, $j \leq m$, $f(j) = i_0$, and $p \in C(i_0)$. Then there exists a chain D which is an open cover of P with $p \in D(j)$ and D following the pattern f in C if and only if for every subcontinuum M of P containing p, there exist a, b with $0 \leq a \leq j \leq b \leq m$ such that $M \subseteq \bigcup_{a \leq k \leq b} C(f(k))$ and $M \cap C(f(k)) \neq \varnothing$ for each $a \leq k \leq b$.*

Many other variations on these results can be located in the papers of Bing,

Lehner, Moise, and the author. The most essential properties are condensed into the following two facts:

(i) every point of the pseudo-arc is an endpoint, and

(ii) for any chain C covering the pseudo-arc and any pattern f whose image has the same number of elements as C, there is a chain D covering the pseudo-arc and following the pattern f in C.

The most common way to construct homeomorphisms of the pseudo-arc is by matching up chains in the domain and range following the same patterns, as indicated in the following theorem.

THEOREM [60]. *Let* $\{C_i\}_{i=1}^{\infty}$ *and* $\{D_i\}_{i=1}^{\infty}$ *be two sequences of chains which are open covers of the chainable continuum* X *such that for each* i:

(1) C_i *and* D_i *have the same number of links*,

(2) $\mathrm{mesh}(C_i) < 1/i$ *and* $\mathrm{mesh}(D_i) < 1/i$, *and*

(3) *there exists a pattern* f_i *such that* C_{i+1} *follows* f_i *in* C_i *and* D_{i+1} *follows* f_i *in* D_i.

Then there exists a homeomorphism $h: X \to\to X$ *such that for each* $p \in X$, *if* $\mathrm{st}(p, C_i) \subset C_i(j) \cup C_i(j+1)$ *then* $h(p) \in D_i(j) \cup D_i(j+1)$.

Bing [6] has shown that pseudo-arcs are very common among continua, at least in the sense of category.

THEOREM [6]. *The collection of all continua homeomorphic to the pseudo-arc which are subcontinua of the Euclidean or Hilbert space* E, *with dimension* $(E) \geq 2$, *forms a dense* G_δ *in the hyperspace of all subcontinua of* E.

A number of nonchainable continua have been constructed with the property that every nondegenerate proper subcontinuum is homeomorphic to the pseudo-arc [6, 25, 38, 96]. Every such continuum is hereditarily indecomposable, almost chainable, almost homogeneous (with each orbit a union of composants), and locally homeomorphic to the pseudo-arc. No such continuum is homogeneous. Many such continua are planar.

3. Characterizations of the pseudo-arc

The first characterization of the pseudo-arc was given by Bing [6]. This has probably been the most important characterization of the pseudo-arc for all subsequent work.

THEOREM [6]. *Every nondegenerate, hereditarily indecomposable, chainable continuum is homeomorphic to the pseudo-arc.*

Bing [10] later showed that the pseudo-arc is characterized by its homogeneity.

THEOREM [10]. *Every nondegenerate, homogeneous, chainable continuum is homeomorphic to the pseudo-arc.*

The result of Moise can also be interpreted as a characterization of the pseudo-arc when combined with a result of Bing.

THEOREM. *Every nondegenerate, hereditarily equivalent, chainable continuum is homeomorphic to either the arc or the pseudo-arc.*

Bing's characterization in terms of homogeneity has been generalized in the following results by J. Charatonik [22] and the author [68], respectively.

THEOREM [22]. *Every nondegenerate chainable continuum which is homogeneous with respect to open mappings is homeomorphic to the pseudo-arc.*

THEOREM [68]. *Every nondegenerate, homogeneous, almost chainable continuum is homeomorphic to the pseudo-arc.*

It is usually straightforward to determine whether a particular continuum is hereditarily indecomposable. Homogeneity is frequently not easy to determine but is taken as an hypothesis. Given a construction of a specific continuum it is usually possible (though sometimes challenging) to determine whether or not the continuum is chainable. However it is much more difficult (and has not always been possible) to determine whether chainability follows from a given combination of properties. The concept of span of a continuum was introduced by Lelek [59], in part to be able to distinguish between chainable and nonchainable continua. It is known that positive span implies that a continuum is nonchainable, and has been used to show the nonchainability of some specific continua [38, 39, 91], but it remains to be determined whether every continuum of zero span is chainable. Oversteegen and Tymchatyn [94] have shown that every continuum of zero span is weakly chainable, i.e., the continuous image of a chainable continuum.

As an attempt to generalize the above characterizations, one can weaken the assumption of chainability or combine homogeneity and hereditary indecomposability.

CONJECTURE. Every nondegenerate, hereditarily indecomposable, weakly chainable continuum is homeomorphic to the pseudo-arc.

CONJECTURE. Every nondegenerate, homogeneous, hereditarily indecomposable continuum is homeomorphic to the pseudo-arc.

One might also ask whether every hereditarily equivalent indecomposable nondegenerate continuum is homeomorphic to the pseudo-arc. In this case [26] as well as for each of the conjectures above [85, 99] it is known that such a continuum must be tree-like. In the case of Bing's classification of the pseudo-arc in terms of hereditary indecomposability, it is known [6] that chainability cannot be replaced by almost chainability, but it is unknown whether it can be replaced by weak chainability. Cook [24] has shown that every map from a continuum onto an hereditarily indecomposable continuum X is confluent. Thus if X is also weakly chainable, it is the confluent image of the pseudo-arc. It is known that both open [103] and monotone [6] maps (subclasses of confluent maps) preserve chainability, while McLean [85] has shown that confluent maps preserve tree-likeness.

Pseudo-arcs play an important role in some areas of homogeneity, and a proof of Conjecture 1 above would represent a major advance in this area. A discussion of the status of work in homogeneity is outside the scope of this article, but can be found in a number of recent survey articles [66, 67, 78, 81, 101, 102]. In Bing's classification of the pseudo-arc in terms of homogeneity, it is clear that chainability cannot be replaced by weak chainability (e.g., the simple closed curve) while the author has shown that chainability can be replaced by almost chainability [68].

4. Mappings properties

The pseudo-arc has many interesting mapping properties, some of which seem unusual in a chainable continuum. Others seem unusual in combination. Several follow from the results in the previous two sections. Bing's results on being able to follow any pattern with an amalgamation of a crooked chain yield the following theorem.

THEOREM. *Every chainable continuum is the continuous image of the pseudo-arc.*

As a corollary, one has that every locally connected continuum is the continuous image of the pseudo-arc. For nondegenerate hereditarily indecomposable continua one always has a map going in the other direction.

THEOREM. *Every nondegenerate hereditarily indecomposable continuum admits a continuous surjection onto the pseudo-arc.*

The author [77] has observed that the first theorem above can be strengthened by imposing additional conditions on the map.

THEOREM [77]. *If X is a weakly chainable nondegenerate continuum and $\varepsilon > 0$, there exists a continuous surjection from the pseudo-arc onto X such that each point inverse has diameter less than ε.*

Under certain classes of maps, however, the pseudo-arc admits maps only onto itself. See also Conjecture 1, and the discussion following it, in the previous section.

THEOREM [6]. *The nondegenerate image of the pseudo-arc under a monotone map is homeomorphic to the pseudo-arc.*

THEOREM [103]. *The nondegenerate image of the pseudo-arc under an open map is homeomorphic to the pseudo-arc.*

Every nondegenerate subcontinuum of the pseudo-arc is homeomorphic to the pseudo-arc. Cornette [28] has shown that every subcontinuum of the pseudo-arc is also a retract of the pseudo-arc. M. Smith [106] has results on the existence of retracts satisfying additional properties.

THEOREM [28]. *Every subcontinuum of the pseudo-arc is a retract of the pseudo-arc.*

THEOREM [106]. *Suppose X is a pseudo-arc, X is irreducible from point P to point Q, and Y is a pseudo-arc which contains X and is the union of two closed point sets H and K so that X is a component of H, $X \cap K = \{Q\}$, and $\mathrm{Bd}(H) = \mathrm{Bd}(K) = H \cap K$. Then there is a retraction h of Y onto X such that $h(K) = Q$, $h^{-1}(P) = P$, and h maps $Y - X$ onto the composant of X containing Q.*

M. Smith [106, 107] has used his retraction result to construct inverse limits with $\aleph_1$ factors, each a pseudo-arc, with each bonding map a retraction, such that the inverse limit is a nonmetric version of the pseudo-arc.

Mackowiak has shown the existence of many retracts onto the pseudo-arc.

THEOREM. *The pseudo-arc is a retract of every hereditarily indecomposable continuum which contains it.*

In response to a question of Bing, Lehner showed that the pseudo-arc satisfies strong homogeneity conditions. This was initially suggested by results of Bing.

THEOREM [5]. *Suppose (P_i, Q_i, R_i) $(i = 1, 2)$ is a triple of points of the pseudo-arc M such that P_i and R_i belong to the same composant of M, and Q_i belongs to a different composant. Then there is a homeomorphism T carrying P_1 to P_2, Q_1 to Q_2, R_1 to R_2, and M onto itself.*

THEOREM [57]. *If $\{A_1, A_2, \ldots, A_n\}$ and $\{B_1, B_2, \ldots, B_n\}$ are finite collections of subcontinua of the pseudo-arc P such that A_i and A_j (respectively, B_i and B_j) are in distinct composants of P if $i \neq j$, then every homeomorphism $h : \bigcup_{i=1}^{n} A_i \to\to \bigcup_{i=1}^{n} B_i$ can be extended to a homeomorphism $\tilde{h}$ of P onto itself.*

THEOREM [57]. *Suppose K_1, K_2 are closed subsets of the pseudo-arc P and each have the same number of components $H_{1,1}, H_{1,2}, \ldots, H_{1,N}$ and $H_{2,1}, H_{2,2}, \ldots, H_{2,N}$ respectively. Suppose also that T is a homeomorphism of K_1 onto K_2. Let m be the minimum of N and 3. Then a necessary and sufficient condition that T can be extended to a homeomorphism of P onto P is that for any m points $P_1, \ldots, P_m$ of K_1 there exists a homeomorphism of P onto P taking P_i to $T(P_i)$ $(i \leq m)$.*

It should be noted that the statement of the last theorem is different from that given by Lehner. He defined m as the maximum of 2 and $N - 1$. When $N = 3$ the author [83] has observed that it is not sufficient to be able to extend the homeomorphism for each pair of points. An ε-version of the extension result of Lehner has recently been given by Kawamura and Prajs [45].

Clearly, composant considerations, for the pseudo-arc as well as for its subcontinua, are necessary when determining whether homeomorphisms between closed subsets can be extended to homeomorphisms of the entire

pseudo-arc. Lehner showed that for closed sets having only a finite number of components these are the only considerations necessary. The author
[83] has shown that in general composant considerations alone are not sufficient and has determined necessary and sufficient considerations, in terms
of the hyperspace of subcontinua, for homeomorphisms between arbitrary
closed subsets to extend to homeomorphisms of the pseudo-arc.

Bing [6] showed that in terms of category the pseudo-arc is very common
among continua. The author [74] has shown that homeomorphisms of the
pseudo-arc are also very common. This result was independently obtained
by M. Smith [105].

THEOREM [74, 105]. *The self-maps of the pseudo-arc which are homeomorphisms onto their images form a dense G_δ in the space of all self-maps of the
pseudo-arc.*

COROLLARY [74, 105]. *Every self-map of the pseudo-arc is a near homeomorphism.*

THEOREM [74]. *The collection of homeomorphisms between pseudo-arcs
forms a dense G_δ in the space of all maps between subcontinua of a Euclidean
space E of dimension at least three.*

One can identify the space of maps between subcontinua with the subspace
of the hyperspace of $E \times E$ consisting of the graphs of the maps. The above
theorem is not true in dimension two.

The author [63] and M. Smith [104] have independently shown that there
are uncountably many distinct embeddings of the pseudo-arc in the plane.
The construction given by Moise [92] produces an embedding of the pseudo-
arc in the plane with exactly two composants arcwise accessible from the
complement. One can achieve any desired number of accessible composants.

THEOREM [63, 104]. *For every positive integer n, as well as for $n = \aleph_0$ and
$2^{\aleph_0}$, there exists an embedding of the pseudo-arc in the plane with exactly n
composants arcwise accessible from the complement.*

THEOREM [63, 104]. *There exists an embedding of the pseudo-arc in the
plane with every accessible point in a different composant.*

Mazurkiewicz [84] has shown that there is no embedding of a nondegenerate indecomposable continuum in the plane for which every composant is
accessible.

THEOREM [84]. *If C is a nondegenerate indecomposable plane continuum,
then the union of the composants of C which contain an accessible point is of
the first category in C.*

Every embedding type is dense in the following sense.

THEOREM. *If C is the collection of all pseudo-arcs in the plane (viewed as a
subspace of the hyperspace of all subcontinua of the plane) and for a particular*

pseudo-arc P in the plane D_P is the collection of all pseudo-arcs in the plane equivalent to P under an orientation preserving homeomorphism of the plane, then D_P is dense in C.

For many particular embeddings of the pseudo-arc in the plane, including the embedding of Moise, it can be shown that only a nowhere dense collection of homeomorphisms extends to homeomorphisms of the plane. It is at present unknown whether this is true for every embedding. At present very little is known about which homeomorphisms of the pseudo-arc extend to homeomorphisms of the plane under which embeddings. Though the homeomorphisms and self-maps of the pseudo-arc exhibit very rich structure, indications are that those which are extendible to homeomorphisms of plane are much more restrictive in their behavior.

The author [82] has shown, however, that homeomorphisms of the pseudo-arc in the plane always extend to homeomorphisms of $\mathbb{R}^3$.

THEOREM [82]. *If A and B are chainable continua in $\mathbb{R}^2$ and $h: A \to B$ is a homeomorphism, then h extends to a homeomorphism $\tilde{h}: \mathbb{R}^3 \to \mathbb{R}^3$.*

In particular, for a pseudo-arc in the plane, its homogeneity can be achieved with homeomorphisms of $\mathbb{R}^3$.

THEOREM [82]. *Every pseudo-arc in the plane is homogeneously embedded in $\mathbb{R}^3$.*

Since the pseudo-arc is chainable, it has the fixed-point property [34]. The author [60] has shown that, despite restrictions imposed by its hereditary indecomposability, the pseudo-arc admits nonidentity stable homeomorphism, thus answering a question of Bing [9]. Kennedy [46] has used the techniques developed by the author to show that his result can be combined with a version of the extension result of Lehner. Clearly any stable homeomorphism must setwise preserve each composant.

THEOREM [60]. *Let U be an open subset of the pseudo-arc P. Let p and q be distinct points of P such that the continuum M irreducible between p and q does not intersect $\mathrm{cl}(U)$. There exists a homeomorphism $h: P \to P$ such that $h(p) = q$ and $h|_U = 1_U$.*

THEOREM [46]. *Let U be an open subset of the pseudo-arc P. Let M be a subcontinuum of P which does not intersect $\mathrm{cl}(U)$. Let $h: M \to M$ be a homeomorphism. There exists a homeomorphism $\tilde{h}: P \to P$ such that $\tilde{h}|_M = h$ and $\tilde{h}|_U = 1_U$.*

It remains to be determined whether the pseudo-arc has the complete invariance property, i.e., whether every nonempty closed subset is the fixed-point set of some self-map of the pseudo-arc.

The author [74] has shown that every self-map of a chainable continuum is semiconjugate to a homeomorphism of the pseudo-arc.

THEOREM [74]. *If* $f: X \to X$ *is a map of the chainable continuum* X, *there exists a homeomorphism* $h: P \to P$ *of the pseudo-arc* P *and a continuous surjection* $\varphi: P \to X$ *such that* $f\varphi = \varphi h$. *If* f *is a surjection then* h *can also be chosen to be a surjection.*

In the above theorem one simultaneously constructs the homeomorphism h and the projection φ. One cannot in general have φ given before one constructs h. In many cases the lift of f to h can be chosen to preserve much of the dynamics of f. Thus the homeomorphisms of the pseudo-arc exhibit most of the dynamics present in maps of the interval, as well as much additional dynamics. One particular application of this construction has been exhibited by Kennedy [47]. This construction seems to offer potential for many more interesting homeomorphisms of the pseudo-arc.

THEOREM [47]. *The tent map on the interval is semiconjugate to homeomorphisms of the pseudo-arc. One such homeomorphism is transitive, while another such homeomorphism has wandering points.*

The author [80] has recently shown that the homeomorphisms of the pseudo-arc which are near the identity contain representations of all behavior present in homeomorphisms of the pseudo-arc.

THEOREM [80]. *Every homeomorphism of the pseudo-arc is conjugate to an* ε-*homeomorphism.*

While much work on the pseudo-arc has treated it as given by a defining sequence of crooked chains, a few results of interest have been obtained using inverse limits. One of the first such was obtained by Henderson [37].

THEOREM [37]. *There exists a map* $f: [0, 1] \to [0, 1]$ *such that the inverse limit obtained by using* f *as each one step bonding map is homeomorphic to the pseudo-arc.*

The dynamics exhibited by the map f of Henderson are elementary, with the endpoints fixed and every other point attracted to 0. Minc and Transue [87] have recently obtained a similar result using a map with more interesting dynamics.

THEOREM [87]. *There exists a map* $f: [0, 1] \to [0, 1]$ *such that* f *is transitive and the inverse limit obtained by using* f *as each one step bonding map is homeomorphic to the pseudo-arc.*

For every homeomorphism of the Cantor set there is an embedding of the Cantor set in the pseudo-arc (e.g., one with each point of the Cantor set in a different composant [23]) such that the homeomorphism extends to a homeomorphism of the pseudo-arc. This observation, made independently by Kennedy [48] and the author [83], yields the following result.

THEOREM [48, 83]. *The pseudo-arc admits homeomorphisms of positive entropy.*

The following result by the author [77] shows that any two pseudo-arcs which are setwise near each other are homeomorphically close.

THEOREM [77]. *If P is a pseudo-arc contained in the metric space X and $\varepsilon > 0$, there exists $\delta > 0$ such that for any pseudo-arc Q contained in X, such that the Hausdorff distance from P to Q is less than δ, there is a homeomorphism $h: P \to Q$ of P onto Q such that $\mathrm{dist}(x, h(x)) < \varepsilon$ for each point $x \in P$.*

Although the homeomorphisms of the pseudo-arc exhibit a very rich structure, and much of the interest in the pseudo-arc stems from the fact that it is hereditarily indecomposable, both of these are lost when one takes products.

THEOREM [40]. *The product of any two nondegenerate continua is aposyndetic, and hence decomposable.*

THEOREM [4]. *If $X = \prod_{i=1}^{n} P_i$ is a product where each P_i is homeomorphic to the pseudo-arc, then every homeomorphism $h: X \to X$ is of the form $h = (h_1, \ldots, h_n)$ where, for some permutation α of $(1, 2, \ldots, n)$, h_i is a homeomorphism of P_i onto $P_{\alpha(i)}$ for each i.*

5. Continuous decompositions

Jones [41] has shown that every decomposable homogeneous plane continuum admits a continuous decomposition into homogeneous, mutually homeomorphic, nonseparating plane continua, with the decomposition space homeomorphic to a simple closed curve. He and Bing [11] have shown that there is a circle-like homogeneous plane continuum with such a continuous decomposition into pseudo-arcs. This continuum is referred to as the *circle of pseudo-arcs* (not to be confused with the *pseudo-circle* [6], which is hereditarily indecomposable and not homogeneous [30, 97]). Every homeomorphism or self-map of the simple closed curve lifts under the projection map to a homeomorphism or self-map of the circle of pseudo-arcs with motion inside individual elements of the decomposition being free. Hagopian [33] and Rogers [98] have shown that a similar result is true for the solenoids. The author [75] has extended this by showing that for each solenoid there is a unique circle-like solenoid of pseudo-arcs, which is also homogeneous. This gives a complete classification of homogeneous, circle-like continua.

THEOREM [75]. *Every homogeneous nondegenerate circle-like continuum is homeomorphic to either a simple closed curve, a solenoid, a circle of pseudo-arcs, a solenoid of pseudo-arcs, or the pseudo-arc. To each solenoid corresponds a unique solenoid of pseudo-arcs obtainable by lifting an inverse system of simple closed curves giving the solenoid to an inverse system of circles of pseudo-arcs.*

The author [76] has shown that a similar construction can be done for any one-dimensional continuum.

THEOREM [76]. *For each one-dimensional continuum M there exists a one-dimensional continuum $\widetilde{M}$ with a continuous decomposition G into pseudo-arcs such that $\widetilde{M}/G$ is homeomorphic to M. If M is planar, circle-like, tree-like, or homogeneous, $\widetilde{M}$ can also be chosen to have such property.*

This yields at least two new classes of homogeneous continua, one corresponding to each of the Menger universal curve [1] and the solenoidal inverse limits of Menger universal curves as described initially by Case [21] and later studied by Minc and Rogers [86]. In the study of indecomposable homogeneous continua, many examples of hereditarily indecomposable continua containing pseudo-arcs as subcontinua have been constructed as candidates for homogeneous continua [6, 25]. One can show that if an hereditarily indecomposable homogeneous continuum contains a pseudo-arc then it admits a continuous decomposition into maximal pseudo-arcs. The quotient space under this decomposition would also be an hereditarily indecomposable homogeneous continuum. If no new pseudo-arcs were introduced in the quotient space, one would have such an example with no nondegenerate chainable subcontinua. The author [71] has shown that such a decomposition does not introduce new chainable continua.

THEOREM [71]. *If X is an hereditarily indecomposable continuum with a continuous decomposition G into pseudo-arcs such that the quotient space X/G is homeomorphic to the pseudo-arc, then X is itself homeomorphic to the pseudo-arc.*

Kawamura and Prajs [45] have shown that the assumption that X is hereditarily indecomposable can be replaced with the assumption that X is atriodic.

The above results concentrate on one-dimensional spaces with continuous decompositions into pseudo-arcs. Walsh and the author [61] have shown that there also exist such decompositions of some two-dimensional spaces.

THEOREM [61]. *There exists a continuous decomposition of the plane into pseudo-arcs.*

Such decompositions also exist for the 2-sphere, the torus, and various other 2-manifolds. Our decomposition has much translational symmetry, but it is unknown in general which homeomorphisms of the plane lift to homeomorphisms of the decomposition which preserve elements of the decomposition. The construction is geometric and straightforward, requiring only a minimal knowledge of the pseudo-arc. Examination of it is recommended to anyone interested in continuous decompositions.

6. Homeomorphism groups

Though there has been much interest in the topological structure of the group of homeomorphisms of the pseudo-arc and related continua, at present very little is known. The author has shown that the group of homeomorphisms of the pseudo-arc contains copies of a variety of compact, zero-dimensional groups.

THEOREM [64]. *For every positive integer* n *there exists a homeomorphism of the pseudo-arc of period* n. *There exists an embedding of the pseudo-arc in the plane such that this homeomorphism is a restriction of a period* n *rotation of the plane.*

THEOREM [73]. *Every inverse limit of finite solvable groups acts effectively on on the pseudo-arc.*

This includes actions by p-adic Cantor groups [69]. It is at present unknown whether every zero-dimensional compact group, or even every finite group, acts effectively on the pseudo-arc. The smallest group for which such an action is not currently known is A_5, the simple group of order 60. The results on periodic homeomorphisms have been extended by Toledo [109, 110] to give the following theorem.

THEOREM [109, 110]. *If* $1 = x_0, x_1, \ldots, x_n$ *is a sequence of positive integers such that, for each* $i < n$, x_i *divides* x_{i+1}, *then there exists a homeomorphism of the pseudo-arc of period* x_n *with points of each order* x_i *and points of no other orders.*

THEOREM [109]. *If* C_0 *is a chainable continuum and* n *is a positive integer, there exists a chainable continuum* C_1 *and a homeomorphism* h *of* C_1 *of period* n *such that the fixed point set of* h *is homeomorphic to* C_0.

The author [70] has shown that the homeomorphism group of the pseudo-arc contains no positive-dimensional compact subsets.

THEOREM [70]. *Every compact subset of the group of homeomorphisms of the pseudo-arc is zero-dimensional.*

The analog of the above result is true for any continuum all of whose sufficiently small subcontinua are pseudo-arcs. It is currently unknown whether the group of homeomorphisms of the pseudo-arc has any nondegenerate connected subsets, though it is expected that it is totally disconnected. It is known that the group of homeomorphisms of the Menger universal curve is totally disconnected. Brechner [13] has, in fact, shown that a special type of separation is possible.

THEROEM [13]. *Suppose* f *and* g *are distinct homeomorphisms of the Menger universal curve* M *onto itself. There exists* $\varepsilon > 0$ *and a separation*

$H(M) = A \cup B$ *of the space of homeomorphisms of* M *such that* $f \in A$, $g \in B$, *and* $\text{dist}(A, B) > \varepsilon$.

The author [74] has shown that such a result is not true for the pseudo-arc.

THEOREM [74]. *Let* f *be a homeomorphism of the pseudo-arc* P *and* $\varepsilon >$ 0. *There exist homeomorphisms* $f_1, f_2, \ldots, f_n$ *such that* $f = f_n \circ f_{n-1} \circ \cdots \circ f_1$ *and, for each* $i \le n$, $\text{dist}(f_i, 1_P) < \varepsilon$.

In each of the above results we are using the sup metric to measure the distance between homeomorphisms.

At present nothing is known about the dimension of the group of homeomorphisms of the pseudo-arc. It is expected that it will eventually be shown to be infinite-dimensional. We also know very little about normal subgroups of the group of homeomorphisms, essentially what follows from the above results or from the mapping properties described in Section 4.

The structure of the group of homeomorphisms of the pseudo-arc is still a fresh area for investigation where the introduction of additional techniques should prove useful.

7. Questions

Below we list some questions concerning the pseudo-arc. One should also see references [16], [72], and [79], as well as the problem section of this book. We have not attempted to identify the source of each question (sometimes complicated to fully determine through all of its variations) or the current state of knowledge concerning each. If any interests the reader, the author will be glad to provide information on what he knows about its status and history, with additional references. Below we only list questions which the author did not submit for inclusion in the separate problem section of this book. P denotes the pseudo-arc.

QUESTION 1. Does the pseudo-arc (or any chainable continuum) admit a pointwise periodic, nonperiodic homeomorphism?

QUESTION 2. Does each (periodic) homeomorphism h of the pseudo-arc have a square root (i.e., a homeomorphism g such that $g^2 = h$)?

QUESTION 3. What is a classification, up to conjugacy, of the periodic homeomorphisms of the pseudo-arc? Is this determined entirely by which prime periods occur and whether the fixed-point set is nondegenerate?

QUESTION 4. Does every compact, zero-dimensional group act effectively on the pseudo-arc?

QUESTION 5. Are the periodic (respectively almost periodic or pointwise periodic) homeomorphisms dense in the group of all homeomorphisms of the pseudo-arc? (Conjecture is no.)

QUESTION 6. Must the homeomorphism group of a homogeneous continuum either contain an arc or be totally disconnected?

QUESTION 7. If G is the collection of homeomorphisms of the pseudo-arc

P which leave every composant invariant, is G dense in the full homeomorphism group of P? of first category?

QUESTION 8. Do minimal normal subgroups of the groups of homeomorphisms characterize chainable continua?

QUESTION 9. What is the structure of the collection of homeomorphisms of the pseudo-arc leaving a given point fixed?

QUESTION 10. Under what conditions can a map $f: K \to P$ from a closed subset K of an hereditarily indecomposable continuum H into the pseudo-arc P be extended to a map $\tilde{f}: H \to P$?

QUESTION 11. If the homeomorphism $f: P \to P$ of the pseudo-arc P has positive entropy, does f have homoclinic orbits?

QUESTION 12. What periodic structure does positive entropy of the homeomorphism $f: P \to P$ imply?

QUESTION 13. If the homeomorphism $f: P \to P$ is transitive, are the periodic points of f dense in P?

QUESTION 14. Does the pseudo-arc minus a point admit a minimal homeomorphism, i.e., a homeomorphism with every orbit dense?

REFERENCES

1. R. D. Anderson, *A characterization of the universal curve and a proof of its homogeneity*, Ann. of Math. (2) **67** (1958), 313–324.

2. ____, *Open mappings of compact continua*, Proc. Nat. Acad. Sci. U.S.A. **42** (1956), 347–349.

3. D. Bellamy and J. Lysko, *Factorwise rigidity of the product of pseudo-arcs*, Topology Proc. **8** (1983), 21–27.

4. D. Bellamy and J. Kennedy, *Factorwise rigidity of products of pseudo-arcs*, Topology Appl. **24** (1986), 197–205.

5. R. H. Bing, *A homogeneous indecomposable plane continuum*, Duke Math. J. **15** (1948), 729–742.

6. ____, *Concerning hereditarily indecomposable continua*, Pacific J. Math. **1** (1951), 43–51.

7. ____, *Snake-like continua*, Duke Math. J. **18** (1951), 653–663.

8. ____, *Higher-dimensional hereditarily indecomposable continua*, Trans. Amer. Math. Soc. **71** (1951), 267–273.

9. ____, *The pseudo-arc*, Summary of Lectures and Seminars, Summer Institute on Set Theoretic Topology (Madison, 1955; revised 1958), 1958, pp. 72–75.

10. ____, *Each homogeneous nondegenerate chainable continuum is a pseudo-arc*, Proc. Amer. Math. Soc. **10** (1959), 345–346.

11. R. H. Bing and F. B. Jones, *Another homogeneous plane continuum*, Trans. Amer. Math. Soc. **90** (1959), 171–192.

12. B. Brechner, *Periodic homeomorphisms on chainable continua*, Fund. Math. **64** (1969), 197–202.

13. ____, *Strongly locally setwise homogeneous continua and their homeomorphism groups*, Trans. Amer. Math. Soc. **154** (1971), 279–288.

14. ____, *On stable homeomorphisms and imbeddings of the pseudo-arc*, Illinois J. Math. **22** (1978), 630–661.

15. ____, *Extendable periodic homeomorphisms on chainable continua*, Houston J. Math. **7** (1981), 327–344.

16. ____, *Questions on homeomorphism groups of chainable and homogeneous continua*, Topology Proc. **7** (1982), 391–393.

17. M. Brown and H. Gluck, *Stable structures on manifolds*, Ann. of Math. **79** (1964), 1–17.

18. C. E. Burgess, *Homogeneous continua*, Summary of Lectures and Seminars, Summer Institute on Set-Theoretic Topology (Madison, 1955; revised 1958), pp. 75–78.

19. ____, *Homogeneous continua which are almost chainable*, Canad. J. Math. **13** (1961), 519-528.

20. ____, *A characterization of homogeneous plane continua that are circularly chainable*, Bull. Amer. Math. Soc. **75** (1969), 1354–1356.

21. J. H. Case, *Another 1-dimensional homogeneous continuum which contains an arc*, Pacific J. Math. **11** (1961), 455–469.

22. J. J. Charatonik, *A characterization of the pseudo-arc*, Bull. Acad. Polon. Sci. **26** (1978), 901–903.

23. H. Cook, *On the most general plane closed point set through which it is possible to pass a pseudo-arc*, Fund Math. **55** (1964), 11–22.

24. ____, *Continua which admit only the identity mapping onto nondegenerate subcontinua*, Fund. Math. **60** (1967), 241–249.

25. ____, *Concerning three questions of Burgess about homogeneous continua*, Colloq. Math. **19** (1958), 241–244.

26. ____, *Tree-likeness of hereditarily equivalent continua*, Fund. Math. **68** (1970), 203–205.

27. ____, *A locally compact, homogeneous metric space which is not bihomogeneous*, Topology Proc. **11** (1986), 25–27.

28. J. Cornette, *Retracts of the pseudo-arc*, Colloq. Math. **19** (1968), 235–239.

29. L. Fearnley, *Characterizations of the continuous images of the pseudo-arc*, Trans. Amer. Math. Soc. **111** (1964), 380–399.

30. ____, *The pseudo-circle is not homogeneous*, Bull. Amer. Math. Soc. **75** (1969), 554–558.

31. J. B. Fugate and T. B. McLean, *Compact groups of homeomorphisms on tree-like continua*, Trans. Amer. Math. Soc. **267** (1981), 609–620.

32. C. L. Hagopian, *Indecomposable homogeneous plane continua are hereditarily indecomposable*, Trans. Amer. Math. Soc. **224** (1976), 339–350.

33. C. L. Hagopian and J. T. Rogers, Jr., *A classification of homogeneous circle-like continua*, Houston J. Math. **3** (1977), 471–474.

34. O. H. Hamilton, *A fixed point theorem for pseudo-arcs and certain other metric continua*, Proc. Amer. Math. Soc. **2** (1951), 173–174.

35. M. Handel, *A pathological area preserving C^∞ diffeomorphism of the plane*, Proc. Amer. Math. Soc. **86** (1982), 163–168.

36. G. W. Henderson, *Proof that every compact decomposable continuum which is topologically equivalent to each of its nondegenerate subcontinua is an arc*, Ann. Math. **27** (1960), 421–428.

37. ____, *The pseudo-arc as an inverse limit with one binding map*, Duke Math. J. **31** (1964), 421–425.

38. W. T. Ingram, *Hereditarily indecomposable tree-like continua*, Fund. Math. **103** (1979), 61–64.

39. ____, *Hereditarily indecomposable tree-like continua. II*, Fund. Math. **111** (1981), 95–106.

40. F. B. Jones, *Concerning non-aposyndetic continua*, Amer. J. Math. **70** (1948), 403–413.

41. ____, *On a certain type of homogeneous plane continuum*, Proc. Amer. Math. Soc. **6** (1955), 735–740.

42. ____, *On homogeneity*, Summary of Lectures and Seminars, Summer Institute on Set-Theoretic Topology (Madison, 1955; revised 1958), pp. 68–70.

43. ____, *Homogeneous plane continua*, Proc. Auburn Topology Conference (Auburn, 1969), pp. 46–56.

44. K. Kawamura, *A mapping property of the pseudo-arc*, Preprint.

45. K. Kawamura and J. Prajs, *Another application of the Effros theorem to the pseudo-arc*, Preprint.

46. J. Kennedy, *Stable extensions of homeomorphisms on the pseudo-arc*, Trans. Amer. Math. Soc. **310** (1988), 167–178.

47. ____, *A transitive homeomorphism on the pseudo-arc which is semiconjugate to the tent map*, Preprint.

48. ____, *Positive entropy homeomorphisms on the pseudo-arc*, Preprint.

49. J. Kennedy and J. T. Rogers, Jr., *Orbits of the pseudo-circle*, Trans. Amer. Math. Soc. **296** (1986), 327–340.

50. B. Knaster, *Un continu dont tout soús-continu est indécomposable*, Fund. Math. **3** (1922), 247–286.

51. J. Krasinkiewicz, *On the hyperspaces of hereditarily indecomposable continua*, Fund. Math. **84** (1974), 175–186.

52. ____, *On the hyperspaces of certain plane continua*, Bull. Acad. Polon. Sci. **23** (1975), 981–983.

53. ____, *Mapping properties of hereditarily indecomposable continua*, Houston J. Math. **8** (1982), 507–516.

54. ____, *Hereditarily indecomposable representatives of shape*, Preprint.

55. P. Krupski, *Short proofs that the hyperspaces of subcontinua of the pseudo-arc and of solenoids of pseudo-arcs are Cantor manifolds*, Preprint.

56. K. Kuratowski, *Topology*, vol. II, Academic Press and PWN, New York and Warsaw, 1968.

57. G. R. Lehner, *Extending homeomorphisms on the pseudo-arc*, Trans. Amer. Math. Soc. **98** (1961), 369–394.

58. A. Lelek, *On weakly chainable continua*, Fund. Math. **51** (1962), 271–282.

59. ____, *Disjoint mappings and the span of spaces*, Fund. Math. **55** (1964), 199–214.

60. W. Lewis, *Stable homeomorphisms of the pseudo-arc*, Canad. J. Math. **31** (1979), 363–374.

61. W. Lewis and J. Walsh, *A continuous decomposition of the plane into pseudo-arcs*, Houston J. Math. **4** (1978), 209–222.

62. W. Lewis, *Monotone maps of hereditarily indecomposable continua*, Proc. Amer. Math. Soc. **75** (1979), 361–364.

63. ____, *Embeddings of the pseudo-arc in E^2*, Pacific J. Math. **93** (1981), 115–120.

64. ____, *Periodic homeomorphisms of chainable continua*, Fund. Math. **117** (1983), 81–84.

65. ____, *Homogeneous tree-like continua*, Proc. Amer. Math. Soc. **82** (1981), 470–472.

66. ____, *Homogeneous curves*, Topology Proc. **9** (1984), 85–98.

67. ____, *Homogeneous hereditarily indecomposable continua*, Topology Proc. **5** (1980), 215–222.

68. ____, *Almost chainable homogeneous continua are chainable*, Houston J. Math. **7** (1981), 373–377.

69. ____, *Homeomorphism groups and homogeneous continua*, Topology Proc. **6** (1981), 335–344.

70. ____, *Pseudo-arcs and connectedness in homeomorphism groups*, Proc. Amer. Math. Soc. **87** (1983), 745–748.

71. ____, *The pseudo-arc of pseudo-arcs is unique*, Houston J. Math. **10** (1984), 227–234.

72. W. Lewis (Editor), *Continuum theory problems*, Topology Proc. **8** (1983), 361–394.

73. W. Lewis, *Compact group actions on chainable continua*, Houston J. Math. **11** (1985), 225–236.

74. ____, *Most maps of the pseudo-arc are homeomorphisms*, Proc. Amer. Math. Soc. **91** (1984), 147–154.

75. ____, *Homogeneous circle-like continua*, Proc. Amer. Math. Soc. **89** (1983), 163–168.

76. ____, *Continuous curves of pseudo-arcs*, Houston J. Math. **11** (1985), 91–99.

77. ____, *Observations on the pseudo-arc*, Topology Proc. **9** (1984), 329–337.

78. ____, *Homogeneous continua and continuous decompositions*, Topology Proc. **8** (1983), 71–84.

79. ____, *Continuum theory problems: Update*, Topology Proc. **9** (1984), 375–382.

80. ____, *Homeomorphisms of the pseudo-arc are essentially small*, Topology Appl. **34** (1990), 203–206.

81. ____, *Classification of homogeneous continua*, Soochow J. Math. (to appear).

82. ____, *Extending homeomorphisms of chainable continua to $\mathbb{R}^3$*, in preparation.

83. ____, *Extending homeomorphisms to the pseudo-arc*, in preparation.

84. S. Mazurkiewicz, *Sur les points accessible des continus indécomposables*, Fund. Math. **14** (1929), 107–115.

85. T. B. McLean, *Confluent images of tree-like curves are tree-like*, Duke Math. J. **39** (1972), 465–473.

86. P. Minc and J. T. Rogers, Jr., *Some new examples of homogeneous curves*, Topology Proc. **10** (1985), 347–356.

87. P. Minc and W. R. R. Transue, *A transitive map on* [0, 1] *whose inverse limit is the pseudo-arc*, Preprint.

88. J. Mioduszewski, *A functional conception of snake-like continua*, Fund. Math. **51** (1962), 179–189.

89. ____, *Everywhere oscillating functions, extension of the uniformization and homogeneity of the pseudo-arc*, Fund. Math. **56** (1964), 131–155.

90. ____, *On pseudocircles*, Theory of Sets and Topology, Deutsch. Verlagwissensch. Berlin, 1972, pp. 363–375.

91. L. Mohler and L. Oversteegen, *A hereditarily indecomposable, hereditarily non-chainable, planar tree-like continuum*, Fund. Math. **122** (1984), 237–246.

92. E. E. Moise, *An indecomposable plane continuum which is homeomorphic to each of its nondegenerate subcontinua*, Trans. Amer. Math. Soc. **63** (1948), 581–594.

93. ____, *A note on the pseudo-arc*, Trans. Amer. Math. Soc. **63** (1949), 57–58.

94. L. G. Oversteegen and E. D. Tymchatyn, *On span and weakly chainable continua*, Fund. Math. **122** (1984), 159–174.

95. J. Prajs, *Homogeneous, tree-like continua are hereditarily indecomposable*, Preprint.

96. P. Roberson, *Hereditarily indecomposable Case-Chamberlin type continua*, Preprint.

97. J. T. Rogers, Jr., *The pseudo-circle is not homogeneous*, Trans. Amer. Math. Soc. **148** (1970), 417–428.

98. ____, *Solenoids of pseudo-arcs*, Houston J. Math. **3** (1977), 531–537.

99. ____, *Homogeneous, hereditarily indecomposable continua are tree-like*, Houston J. Math. **8** (1982), 421–428.

100. ____, *An aposyndetic, homogeneous curve that is not locally connected*, Houston J. Math. **9** (1983), 433–440.

101. ____, *Homogeneous continua*, Topology Proc. **8** (1983), 213–233.

102. ____, *Classifying homogeneous continua*, Preprint.

103. I. Rosenholtz, *Open maps of chainable continua*, Proc. Amer. Math. Soc. **42** (1974), 258–264.

104. M. Smith, *Plane indecomposable continua, prime ends, and embeddings of the pseudo-arc*, Topology Proc. **3** (1978), 295–300.

105. ____, *Every mapping of the pseudo-arc onto itself is a near homeomorphism*, Proc. Amer. Math. Soc. **91** (1984), 163–166.

106. ____, *A hereditarily indecomposable Hausdorff continuum with exactly two composants*, Topology Proc. **9** (1984), 123–143.

107. ____, *On non-metric pseudo-arcs*, Topology Proc. **10** (1985), 385–397.

108. M. Smith and S. Young, *Periodic homeomorphisms on T-like continua*, Fund. Math. **104** (1978), 221–224.

109. J. Toledo, *Finite and compact actions on chainable and tree-like continua*, Ph.D. Dissertation, University of Florida, Gainesville, 1982.

110. ____, *Inducible periodic homeomorphisms of tree-like continua*, Trans. Amer. Math. Soc. **282** (1984), 77–108.

TEXAS TECH UNIVERSITY, LUBBOCK, TEXAS 79409-1042

Contemporary Mathematics
Volume **117**, 1991

Dynamical Properties of the Shift Map
on the Inverse Limit Space

SHIHAI LI

Let X be a compact metric space and let $f : X \to X$ be continuous.

In this paper, we obtain some dynamical properties of the shift map on the inverse limit space. The inverse limit space of the sequence

$$X \xleftarrow{f} X \xleftarrow{f} X \xleftarrow{f} \cdots$$

is defined to be the set of points $\mathbf{x} = (x_0, x_1, x_2, \ldots)$ satisfying $f(x_{i+1}) = x_i$ with the metric

$$\mathbf{d}(\mathbf{x}, \mathbf{y}) = \sum_{i=0}^{\infty} \frac{d(x_i, y_i)}{2^i},$$

where d is a metric on X. Let $\varprojlim (X, f)$ denote the inverse limit space. The shift map $\sigma_f : \varprojlim (X, f) \to \varprojlim (X, f)$ is defined by $\sigma_f((x_0, x_1, \ldots)) = (f(x_0), x_0, x_1, \ldots)$. Obviously, σ_f is a homeomorphism and

$$\sigma_f^{-1}((x_0, x_1, x_2, \ldots)) = (x_1, x_2, \ldots).$$

We will denote σ_f by σ simply. The projection maps $\pi_i : \varprojlim (X, f) \to X$ are defined by $\pi_i((x_0, x_1, \ldots, x_i, \ldots)) = x_i$ for $i = 0, 1, \ldots$. They are continuous.

An orbit of a point x is defined by $\mathrm{Orb}(x) = \{y : y = f^n(x)$ for $n \geq 0\}$. A point x is called a *periodic point* if $x = f^n(x)$ for some positive integer n. The smallest such integer is called the period of x. A point is called an *ω-limit point* if it is an accumulation point of the forward orbit of some point in X. A point is called a *nonwandering point* if any neighborhood of this point intersects the image of the neighborhood under f^n for some $n > 0$. A point is called a *recurrent point* if it is an ω-limit point of itself. A point is called an *almost periodic point* if, for any neighborhood

---

1980 *Mathematics Subject Classification* (1985 *Revision*). Primary 58F13.

The detailed version of this paper will be submitted for publication elsewhere. No proof is provided here.

U, $\{n \geq 0; f^n(s) \in U\}$ has bounded gaps, i.e., if we write $\{n > 0; f^n(s) \in U\}$ as $\{0 = n_0 < n_1 < n_2 < \cdots\}$, then $\{n_{k+1} - n_k\}$ is bounded. A point x is called a *chain recurrent point* if for any number $\varepsilon > 0$ there exists a sequence of points, $x_0 = x, x_1, \ldots, x_{n-1}, x_n = x$, such that $|f(x_i) - x_{i+1}| < \varepsilon$ for $i = 0, 1, \ldots, n-1$.

The collection of the chain recurrent points, nonwandering points, recurrent points, almost periodic points and periodic points are denoted, respectively, by $\mathrm{CR}(f)$, $\Omega(f)$, $\mathrm{R}(f)$, $\mathrm{AP}(f)$, and $\mathrm{P}(f)$, or CR, Ω, R, AP, and P. The collection of the ω-limit points of one point x is denoted by $\omega(x, f)$ or $\omega(x)$.

THEOREM A. *For any continuous map, the following properties hold.*

(1) *Let* $x \in X$ *and let* $\mathbf{x} \in \varprojlim(X, f)$ *satisfy* $\mathbf{x} = (x, x_1, x_2, \ldots)$; *then*
$$\omega(\mathbf{x}, \sigma) = \varprojlim(\omega(x, f), f);$$

(2) $\mathrm{CR}(\sigma) = \varprojlim(\mathrm{CR}(f), f)$;

(3) $\mathrm{R}(\sigma) = \varprojlim(\mathrm{R}(f), f)$;

(4) $\mathrm{AP}(\sigma) = \varprojlim(\mathrm{AP}(f), f)$;

(5) *When* f *is onto,* $\Omega(\sigma) = \varprojlim(\Omega(f), f)$.

We give a counterexample to show that (5) is not true when f is not onto.

A subset M in X is called a minimal set if M is a nonempty closed invariant set and M has no proper nonempty closed invariant subset. It is equivalent to say that every point in M has a dense orbit in M. We have the following corollary.

COROLLARY 1. *If a subset* M *in* X *is a minimal set, then* $\varprojlim(M, f)$ *is a minimal set. If a subset* $\widetilde{M}$ *of* $\varprojlim(X, f)$ *is a minimal set, then* $\pi_0(\widetilde{M})$ *is a minimal set in* X.

DEFINITION 1. We say that S is an *ω-scrambled set* if, for any $x, y \in S$ with $x \neq y$,

(1) $\omega(x, f) \backslash \omega(y, f)$ is uncountable;

(2) $\omega(x, f) \cap \omega(y, f)$ is nonempty;

(3) $\omega(x, f)$ is not contained in the set of periodic points.

If S is uncountable, then we say that f has *ω-chaos* or f is *ω-chaotic*.

THEOREM B. *If* f *is* ω-chaotic, then $\sigma : \varprojlim(X, f) \to \varprojlim(X, f)$ *is* ω-chaotic.

We can not prove the converse of this theorem. But we can prove that if we substitute the first condition of Definition 1 by the condition

(*) $\omega(x) \backslash \omega(y)$ contains an infinite minimal set,

then the converse of Theorem B is true. This is given as a corollary below. If a map satisfies the condition (*) and also the second and third conditions of ω-chaos, then we say that this map is *ω-chaotic with condition* (*).

COROLLARY 2. *f is ω-chaotic with condition (*) iff σ_f is ω-chaotic with condition (*).*

On the interval, f is ω-chaotic iff f is ω-chaotic with condition (*).

DEFINITION 2. A continuous map $f : X \to X$ is called *topologically transitive* if for any pair of open sets $U, V \subset X$ there exists $k > 0$ such that $f^k(U) \cap V \neq \varnothing$.

DEFINITION 3. A continuous map $f : X \to X$ has *sensitive dependence on initial conditions* if there exists $\delta > 0$ such that, for any $x \in X$ and any neighborhood N of x, there exists $y \in N$ and $n \geq 0$ such that $d(f^n(x), f^n(y)) > \delta$.

DEFINITION 4. Let $f : X \to X$ be a continuous map. f is said to be *chaotic* on X in the sense of Devaney if there is a closed invariant set $D \subset X$ such that the following conditions are satisfied.

(1) $f|_D$ is topologically transitive.
(2) $f|_D$ has sensitive dependence on the initial conditions.
(3) The periodic points of f in D are dense in D.

THEOREM C. *σ_f is chaotic in the sense of Devaney iff f is.*

COROLLARY 3. *Let $f : I \to I$ be a continuous map on the interval. The following are equivalent.*

(1) *σ_f is chaotic.*
(2) *σ_f is ω-chaotic.*
(3) *f has positive topological entropy.*

Our final result concerns branched 1-manifolds studied by Williams [W]. The definition of a branched 1-manifold is analogous to the definition of a 1-manifold. The only difference is that for a branched 1-manifold there are three types of coordinate neighborhoods. These are the real line R, $H = \{x \in R : x \leq 0\}$ and $Y = \{(x, y) \in R^2 : y = 0 \text{ or } y = \phi(x)\}$. Here $\phi : R \to R$ is a fixed C^∞ function such that $\phi(x) = 0$ for $x \leq 0$ and $\phi(x) > 0$ for $x > 0$. Let K be a branched 1-manifold. The tangent bundle of K, $T(K)$, may be defined. A differential map $g : K_1 \to K_2$ of branched 1-manifolds induces a map $Dg : T(K_1) \to T(K_2)$ of their tangent bundles. g is an *immersion* if Dg is a monomorphism on the tangent space at each point.

DEFINITION 5. If $g : K \to K$ is a C^r immersion of a branched C^r manifold $(r \geq 1)$, then g is *expanding* relative to a Riemannian metric $\| \ \|$ on $T(K)$, if there are constants $C > 0$, $\lambda > 1$, such that

$$\|(Df)^n(v)\| \geq C\lambda^n \|v\|$$

for all $n \in \mathbf{Z}^+$, $v \in T(K)$.

THEOREM D. *Let K be a compact branched 1-manifold. Suppose f is expanding and $\Omega(f) = K$. Then f and σ_f are ω-chaotic and chaotic.*

At last we give a method of constructing chaotic homeomorphisms on chainable continua, especially on pseudo-arcs. If one can find a transitive map f on the interval $[0, 1]$, then the shift map σ_f is ω-chaotic and chaotic in the sense of Devaney. Since Minc and Transue [MT] have constructed a transitive map on $[0, 1]$ whose inverse limit space is a pseudo-arc, we get an ω-chaotic and chaotic homeomorphism in the sense of Devaney on a pseudo-arc. Note that a transitive map is an onto map. The chaotic set must be the whole space. The word "chaotic" is the same as in [K] where Kennedy constructed a chaotic homeomorphism on a pseudo-arc by a different method.

Throughout this work, I have had lots of help from Professor L. Block. I would like to thank him very much. I began this work after a discussion with Professor J. Mayer on the way back from the Spring Topology Conference at Knoxville, Tennessee, in 1989. In fact, Theorem B and D are answers to his questions. I would like to thank him very much for the valuable discussion. I would also like to express my appreciation to Professor B. Brechner and Professor J. Keesling who have helped me learn about the inverse limit space and continuum theory.

REFERENCES

[B] L. Block, *Diffeomorphisms obtained from endomorphisms*, Trans. Amer. Math. Soc. **214** (1975), 403–413.

[K] J. Kennedy, *The construction of chaotic homeomorphisms on chainable continua*, preprint.

[L] S.-H. Li, *ω-chaos and topology entropy*, Preprint.

[M] M. Misiurewicz, *Horseshoes for continuous mappings of an interval*, Bull. Acad. Pol. Sci. **27** (1979), 167–169.

[MT] P. Minc and W. R. R. Transue, *A transitive map on* $[0, 1]$ *whose inverse limit is the pseudo-arc*, preprint.

[W] R. F. Williams, *One-dimensional non-wandering sets*, Topology **6** (1967), 473–487.

DEPARTMENT OF MATHEMATICS, UNIVERSITY OF FLORIDA, GAINESVILLE, FLORIDA 32611
E-mail address: shi@math.ufl.edu

Contemporary Mathematics
Volume **117**, 1991

Minimal Sets, Wandering Domains, and Rigidity in the 2-Torus

ALEC NORTON

1. Introduction. Dimension one

The following theorem has been an influential model for rigidity in geometric dynamics.

DENJOY'S THEOREM (1932). *If $f: S^1 \to S^1$ is a C^2 diffeomorphism with no periodic points, then f is topologically conjugate to rotation by an irrational angle.*

This theorem represents a kind of topological rigidity that takes place on the circle: at a certain minimum smoothness (and, in fact, $C^{1+\text{b.v.}}$ is Denjoy's actual hypothesis), some dynamical condition (no periodic points) determines the dynamics (up to topological conjugacy). Rigidity in this case means that behaviors available in the C^1 category are foreclosed in the smoother category: the dynamics becomes resistant to distortion within a suitable space of diffeomorphisms.

It is the aim of this discussion to address this situation in two dimensions. Are there corresponding smoothness constraints analogous to this for surface diffeomorphisms? In particular, is there a theorem analogous to Denjoy's for diffeomorphisms of the 2-torus? (See Theorems A, B, and C below.)

For the moment we will defer a precise formulation of these questions. For now, we consider Denjoy's Theorem in a little more detail. To review some basic facts, recall that if f is a homeomorphism of the circle, the most important conjugacy invariant is called the rotation number

$$\rho(f) = \lim_{n \to \infty} \frac{F^n(x)}{n} \bmod 1$$

1991 *Mathematics Subject Classification.* Primary 58F08.
Partially supported by an NSF Grant.
The final version of this paper will be submitted for publication elsewhere.

which is independent of the lift F of f to $\mathbb{R}$ and independent of the choice of x. Poincare showed that f has rational rotation number p/q in lowest terms if and only if f has a periodic orbit of period q. If f has no periodic orbits, then f is always semiconjugate to R_α = irrational rotation by angle $\alpha = \rho(f)$, i.e., there is a continuous monotone map of the circle onto itself such that $hf = R_\alpha h$. When h is a homeomorphism, f is by definition conjugate to R_α. Otherwise the inverse image under h of each point of some R_α-orbit is a nontrivial interval, and f permutes these intervals in the same order that R_α permutes points. This can happen for up to countably many different R_α-orbits [**Ma**]. The interior of each of these intervals *wanders*, i.e., never meets itself under iteration by f. The complement of all these interiors is a Cantor set which is *minimal* for f, i.e., is a closed invariant set for f containing no closed invariant nonempty proper subsets. (When h is a homeomorphism, the whole circle is minimal since every orbit is dense.)

Denjoy noticed that even for C^1 diffeomorphisms h need not be a homeomorphism, that is, f can have wandering intervals, by constructing examples, called *Denjoy counterexamples*. Of course the point of the theorem is that this behavior is ruled out when f is smoother. Another way of thinking about this is that C^1 diffeomorphisms of the circle can have Cantor minimal sets, but C^2 diffeomorphisms cannot.

There are many proofs of Denjoy's theorem ([**D**], [**He**], [**S**], [**M**], [**KO**], for example), which we will not reproduce. Instead, the following simpler discussion may provide some basic intuition about how the smoothness can play a role in the dynamics, and will shed light on the discussion about two dimensions, in §II.

The simplest construction of a Denjoy counterexample (see, e.g., [**Ni**]) is to specify the derivative of the map f to be equal to one on the minimal Cantor set C (and then to prescribe the derivative on each of a dense collection of intervals so that integrating this derivative yields the required map). We can see why such a diffeomorphism cannot be C^2 via two applications of the Mean Value Theorem.

Suppose for contradiction that f is C^2. (In fact the argument below only requires that f have bounded second derivative.) Since $Df = 1$ on C, we must have $D^2 f = 0$ on C.

Let l_n be the length of $I_n = f^n(I_0)$, where I_0 is some initial interval complementary to the minimal set. The Mean Value Theorem says that there is x_n in I_n such that $Df(x_n) = l_{n+1}/l_n$. Applying the MVT again to Df yields a y_n in I_n such that

$$D^2 f(y_n) \geq \frac{1 - (l_{n+1}/l_n)}{l_n}.$$

By continuity of $D^2 f$, this must tend to zero as $n \to \infty$. Since the lengths

l_n are summable, this means that

$$\sum \left(1 - \frac{l_{n+1}}{l_n}\right) < \infty$$

and hence, by a standard lemma,

$$\prod \frac{l_{n+1}}{l_n} > 0.$$

Since this product telescopes, this implies that the numbers l_n do not tend to zero, a contradiction.

It is this argument which motivates the proof of a two-dimensional analogue (Theorem B) of this result, mentioned in §IV of this paper.

ACKNOWLEDGEMENTS. In one dimension essentially this argument has been known by J. Harrison for many years; it was also used in this form for the two-dimensional result [**Ha**] of G. R. Hall. Harrison has also proved unpublished "no-wandering-domains" results in two dimensions which overlap with the work described below.

II. The two-dimensional case

We now turn to the setting of interest here: diffeomorphisms in two dimensions. In analogy with one dimension, we wish to study the case of diffeomorphisms of the 2-torus $T^2 = \mathbb{R}^2/\mathbb{Z}^2$ which are semiconjugate to an irrational translation $R_{\alpha, \beta}(x) = x + (\alpha, \beta)$, where α and β are chosen so that 1, α and β are linearly independent over the rationals (so that $R = R_{\alpha, \beta}$ is minimal and (uniquely) ergodic on T^2). (Here semiconjugacy means that $hf = Rh$ for some continuous map of the 2-torus onto itself. If h is a homeomorphism then f is (*topologically*) *conjugate* to R.)

Our goal is to find Denjoy-type theorems: if such a map is smooth enough, it must in fact be conjugate to R, under certain further hypotheses (see below). Eventually one would like to eliminate the extra hypotheses, but this is not yet close to being accomplished.

To illustrate the scope of current ignorance on this subject, we mention here a question posed by C. Tresser, which is apparently open:

QUESTION 1. Does there exist any example of a C^2 diffeomorphism of the 2-torus which is semi-conjugate but not conjugate to R?

Note that the product of a Denjoy counterexample with an irrational rotation of the circle provides a C^1 example.

We wish to focus on a particular class of diffeomorphisms as identified in the following

DEFINITIONS. If h is a continuous map of T^2 onto itself, we say that $x \in T^2$ is a *trivial value* of h if $h^{-1}(\{x\})$ is a singleton.

A diffeomorphism f of T^2 is of *Denjoy type* if f is semiconjugate to some ergodic translation R of T^2 by a map h having at most countably many nontrivial values.

f is of *Sierpinski type* if it is of Denjoy type, and furthermore every nontrivial h-inverse image of a point is homeomorphic to the closed unit disc in $\mathbb{R}^2$.

We state our main theorems for diffeomorphisms of Sierpinski type, but it seems likely that a great deal of what we prove here is also true in the more general case of Denjoy type. Precise statements in this more general case will appear elsewhere.

Note that by the relation $hf = Rh$, if $h^{-1}(x)$ has interior, then so does the inverse image of any point in the entire R-orbit of x. Only countably many points of the torus can have this property; in this case h has an inverse defined almost everywhere. The interiors of the nontrivial h-inverse images constitute a collection of wandering domains for f.

When f is of Sierpinski type and h is not a homeomorphism, one model for the complement of the wandering domains is the complement of a dense collection of disjoint round discs—i.e., a Sierpinski curve (hence this terminology). However the diameters of the wandering domains need not *a priori* tend to zero under iteration by f, hence the complement could fail to be locally connected.

We introduce the following notation. For any diffeomorphism of Denjoy type relative to a semiconjugacy h, let $\mathscr{F}$ be the union of the interiors of the sets $h^{-1}(x)$ taken over all nontrivial values x of h. Let $\Gamma = T^2 - \mathscr{F}$. Γ is a closed f-invariant set.

PROPOSITION. *Let f be of Sierpinski type, relative to a semiconjugacy h. Then Γ is connected and is the unique minimal set for f. In particular Γ does not depend on h. Furthermore f is uniquely ergodic.*

PROOF. Γ is connected since it is the complement of a collection of disjoint open discs. For unique ergodicity, we let μ, ν be two invariant probability measures for f. Let N be the (countable) set of nontrivial values for h. Now $\mu(h^{-1}(\text{point})) = 0$ since μ is f-invariant. Hence $\mu(h^{-1}N) = 0$ since N is countable, so that $T^2 - h^{-1}N$ has full μ-measure. In other words, h is almost everywhere injective. This means that

$$h_*^{-1}h_*\mu = \mu.$$

The same is true for ν. But $h_*\mu$ and $h_*\nu$ are R-invariant probability measures, hence are equal. Therefore $\mu = \nu$.

Since any minimal set supports an invariant probability measure, we may deduce immediately that f has a unique minimal set M. By the semiconjugacy relation, one can easily see that every orbit in the complement S of $h^{-1}N$ is dense in S, so that $S \subset M$. By definition of Sierpinski type, each wandering disc is the interior of its closure. This means that Γ is equal to the closure of S. Since Γ is invariant, $\Gamma = M$. This completes the proof.

With care one can construct in a straightforward way a *homeo*morphism of T^2 of Sierpinski type, with wandering round discs. This provides an example of a nontrivial minimal set for T^2 analogous to the Cantor minimal set for a Denjoy counterexample on T^1. The following question has stood open for some time:

QUESTION 2. What are the possible minimal sets for homeomorphisms of T^2?

We will focus instead on

QUESTION 3. Are there diffeomorphisms of Denjoy type?

Currently not even a C^1 example is available, though it seems likely that one exists. (Added in proof: a C^2 example is now available—see note added at the end of this paper.) The remainder of this paper is devoted to ruling out examples in the presence of stronger differentiability hypotheses. To do so, we will have to first discuss some ideas in geometric calculus of several variables in order to set the stage for the theorems to follow.

III. Some results from geometric analysis

For $s \geq 0$, let $\mathscr{H}^s$ denote the Hausdorff (outer) measure in dimension s. This is defined on any metric space by

$$\mathscr{H}^s(A) = \liminf_{\varepsilon \to 0} \sum_{\beta \in \Omega} |B|^s$$

where the infimum is taken over all countable decompositions Ω of A, each of whose elements have diameter less than ε. (Vertical bars denote diameter.) On $\mathbb{R}^n$, $\mathscr{H}^n$ is equivalent to Lebesgue measure. The *Hausdorff dimension* of A is the infimum of the set of s such that $\mathscr{H}^s(A) = 0$. This agrees with topological dimension when A is a smooth submanifold. For further reference see, e.g., [HW, F].

The *box dimension* is a somewhat simpler concept, defined by

$$\mathrm{BD}(A) = \limsup_{\varepsilon \to 0} \frac{\log N(\varepsilon)}{\log(1/\varepsilon)}$$

where $N(\varepsilon)$ is the minimum number of ε-balls required to cover A. It turns out that Hausdorff dimension is always less than or equal to box dimension, and they are equal in many cases, e.g., for self-similar sets.

DEFINITIONS. Let A be a connected set (in, say, some finite-dimensional smooth manifold). Say [N2] that A has *(geometric) connectivity* s, $s \geq 1$, if for every x, y in A, there is a connected subset $S_{x,y}$ of A containing x and y and such that $S_{x,y}$ is s-sigmafinite. (A set is s-sigmafinite if it is a countable union of sets with finite $\mathscr{H}^s$-measure.) We write $A \in \mathrm{GC}(s)$.

Note that $\mathrm{GC}(s) \subset \mathrm{GC}(t)$ for $s < t$, since every s-sigmafinite set is also t-sigmafinite.

Note further that, for $0 \leq s < 1$, $\mathrm{GC}(s)$ contains only singletons, so we neglect this trivial case in our definition. (This is because a connected set larger than a singleton must have topological, and therefore Hausdorff, dimension at least one.) Also every connected set in an n-dimensional manifold is contained in $\mathrm{GC}(n)$. Smooth curves and open sets have connectivity one. Fractal curves, e.g., the von Koch snowflake, do not. Roughly, geometric connectivity is intended to measure how hard it is for pairs of points of A to "see" each other inside A.

Now let g be a C^1 real-valued function defined on a smooth manifold and suppose that $Dg \equiv \text{grad} g$ is zero on A—i.e., suppose that A is a critical set for g. If A is a smooth curve or an open ball, then g must be constant on A by integration. However, in 1935 H. Whitney discovered examples for which A is a (unrectifiable) curve and g is NOT constant on A. (See [W, N2].)

DEFINITION. For $k \in \mathbb{Z}^+$ and $\alpha \in [0, 1)$, say that a function g is of class $C^{k+\alpha+}$ if it is of class C^k and if $|D^k g(x) - D^k g(y)|/(|x - y|^\alpha)$ tends to zero as $|x - y| \to 0$. Note for example that $C^{k+\alpha} \supset C^{k+\alpha+} \supset C^{k+\beta}$ for all $\beta > \alpha$, and that when $\alpha = 0$, $C^{k+\alpha+} = C^{k+\alpha} = C^k$.

We are ready for

THEOREM 1 [N1]. *If A is a critical set for a function $g \in C^1(\mathbb{R}^n, \mathbb{R})$, then*

(a) *if g is C^n, then $g(A)$ has (Lebesgue) measure zero;*

(b) *if A is 1-sigmafinite, then $g(A)$ has measure zero; and*

(c) *if g is $C^{k+\alpha+}$ and A is $(k+\alpha)$-sigmafinite, then $g(A)$ has measure zero.*

It so happens that there are counterexamples to part (c) if $C^{k+\alpha+}$ is weakened to $C^{k+\alpha}$; these are the examples of Whitney. In fact any fractal quasi-circle supports such a sharp counterexample [N2].

Part (a) is a corollary of the ordinary Morse-Sard theorem. One way of putting the theorem in that case is that *the values of a C^n function on a connected set in $\mathbb{R}^n$ are determined (up to a constant) by the values of its derivative on the set.*

This bears a close relation to integration. For example, if γ is a smooth curve along which $Dg = 0$, then $x, y \in \gamma$, $g(x) - g(y)$ is equal, by the Fundamental Theorem of Calculus, to the line integral of Dg along the subarc of γ joining x to y. Since $Dg = 0$, this means that g is constant on γ. While Theorem 1 is a way of obtaining the same result for curves of higher Hausdorff dimension, this integration argument can actually be carried out in more general situations as well, in view of the following theorem.

THEOREM 2 [HN] (Rough Statement). *If ω is a 1-form on $\mathbb{R}^2$ of class C^α, $\alpha \in (0, 1)$, and γ is a Jordan arc with $\text{BD}(\gamma) < 1 + \alpha$, then the integral of ω over γ is well-defined (as a limit of integrals over curves well-approximating γ in a sense not to be elucidated here), and has all the usual properties of the classical line integral.*

In particular the FTC holds for this integral, so that the above integration argument is not so special to smooth or even rectifiable curves as might at first be thought.

IV. Back to toral diffeomorphisms

Let us now reap what we have sown.

PROPOSITION. *If* f *is a* C^1*-diffeomorphism of Sierpinski type, then*

$$\int \log \det Df \, d\mu = 0$$

where μ *is the unique invariant measure for* f.

SKETCH OF PROOF (see [He]). f is uniquely ergodic by the proposition of §II. As a consequence, for any continuous function φ on T^2, the Birkhoff sum

$$\frac{1}{n} \sum_{i=0}^{n-1} \varphi \circ f^i \to \int \varphi \, d\mu$$

converges *uniformly* on T^2 [CFS]. Letting $\varphi = \log \det Df$, the left-hand side above becomes $(\log \det Df^n)/n$ by the chain rule. Now $\log \det Df^n$ cannot converge uniformly to either $+\infty$ or $-\infty$ since $\det Df^n$ must integrate to one by the change of variable formula. The only way out is for the right-hand side to be zero.

This proposition says that on average the Jacobian of f is equal to one on the minimal set Γ for f. In analogy with the construction of Denjoy counterexamples, let us suppose that we have a diffeomorphism of Denjoy type such that $Df = $ the identity matrix at every point of Γ. Theorem A provides a strong form of rigidity in this case.

THEOREM A. *Let* f *be a* C^1*-diffeomorphism of* T^2 *of Sierpinski type (relative to the translation* R*).*

Suppose that $Df = \mathrm{Id}$ *at each point of the minimal set* Γ. *Then* $f = R$ *if any of the following conditions is satisfied:*

(a) f *is* C^2,

(b) $\Gamma \in \mathrm{GC}(1)$, *or*

(c) f *is* $C^{1+\alpha+}$ *and* $\Gamma \in \mathrm{GC}(1 + \alpha)$ *for some* $\alpha \in (0, 1)$.

REMARKS. (i) If, for some $s \geq 1$, the boundary of each wandering disc is in $\mathrm{GC}(s)$, then $\Gamma \in \mathrm{GC}(s)$ also. (To see this, start with a straight line segment connecting any two points of Γ. Replace any subsegment in the interior of a complementary disc with a coterminal boundary arc. The resulting curve is s-sigmafinite.)

In particular there is no C^1 example with a wandering orbit of round discs in case $Df|_\Gamma = \mathrm{Id}$.

(ii) (a) and (b) mean that there is only a narrow window of possibility for such an f: f cannot be C^2, nor can the wandering domains have rectifiable boundaries, but must be fractal sets. Furthermore the theorem stipulates exactly how smooth f can be given a fixed geometry for these boundaries (and hence for Γ).

PROOF OF THEOREM A.

This is an application of Theorem 1. (The proof of part (a) requires only the classical Morse-Sard theorem.)

If f is conjugate to R, then Γ is the whole torus so $Df = \text{Id}$ everywhere, and hence $f = R$. Suppose then that f is not conjugate to R; we argue by contradiction.

Let $g: \mathbb{R}^2 \to \mathbb{R}$ be a component of $F - \text{Id}$, where F is a lift of f to the plane. Since $Df = \text{Id}$ on Γ, we have

$$Dg = 0 \quad \text{on } \Gamma'$$

where Γ' is the lift of Γ to the plane.

In case (a), if f is of class C^2, so is g, so $g(\Gamma')$ has measure zero by Theorem 1. But since $g(\Gamma')$ is a connected subset of $\mathbb{R}$ (hence an interval), g must be constant on Γ'.

The same conclusion can be drawn in the other two cases. For example, in case (c), if $\Gamma \in GC(1 + \alpha)$, then given $x, y \in \Gamma$, there is a connected $(1 + \alpha)$-sigmafinite subset $S_{x,y}$ containing x and y. Then by part (c) of Theorem 1, g is constant on the lift of $S_{x,y}$. Since this holds for every pair of points in Γ, g must be constant on Γ'.

Since the components of $F - \text{Id}$ are constant on Γ', f acts on Γ by translation. In particular f translates the boundary of each wandering domain. This is our contradiction since the domains in the complement of Γ must have areas tending to zero under iteration by f because the torus has finite total area.   Q.E.D.

It is instructive to note that one can instead use a more intuitive integration argument by making use of Theorem 2 instead of Theorem 1. By performing the line integral of the gradient of a component of $F - \text{Id}$ along each boundary, one easily argues via the Fundamental Theorem of Calculus that f restricted to the disc boundaries is a translation and obtains the same contradiction as above. This works fine in the case of smooth curves, or even rectifiable curves. The strength of Theorem 2 is used to get the conclusion in the fractal case. (Note however that this method applies only to Jordan curve boundaries.)

Theorem A has sobering implications for those who wish to construct a C^1 diffeomorphism of Sierpinski type. The wandering domains will not be smooth and the map will not be C^2 if the derivative is to be controlled to the extent of being the identity on the minimal set. However, this derivative restriction is a severe one and the following theorems allow weaker hypotheses on the values of the derivative.

THEOREM 3 [**Ha, N3**]. *If f is a C^3 diffeomorphism of a complete Riemannian surface M with injectivity radius bounded below, and Δ is an open disc in M such that the Jacobian $J(f)$ of f satisfies $J(f) \geq 1$ on $\partial(f^n(\Delta))$ for* *all* $n \geq 1$, *then*

$$\sum_{n=1}^{\infty} \text{area}(f^n(\Delta)) = +\infty.$$

Since T^2 has finite area we have, as a direct corollary,

THEOREM B. *If* f *is a* C^3 *diffeomorphism of* T^2 *of Sierpinski type, and* $J(f) \geq 1$ *on* Γ, *then* $\Gamma = T^2$ *and* f *is conjugate to* R.

IDEA OF PROOF OF THEOREM 3. Careful application of the second-order Taylor's Theorem with remainder to the function $J(f)$. This is analogous to the easier case in one dimension discussed in §1. (Note that the Mean Value Theorem is just the first-order Taylor theorem.) See [N3] for details.

Note that both Theorems A and B have worked by ruling out wandering domains, without making use of the recurrence of f. The proof of the following theorem, which implies Theorem A, makes use of the density of the collection of wandering domains.

THEOREM C [NV]. *Let* f *be a* C^1 *diffeomorphism of Sierpinski type (relative to the ergodic translation* R).

Suppose that for every component Δ *of the complement of* Γ *there is a conformal map* g_Δ *defined on* $\text{cl}(\Delta)$ *such that*

$$Df(y) = Dg_\Delta(y) \quad \text{for all } y \in \partial\Delta.$$

Then $f = R$ *if any of the following conditions are satisfied:*
  (a) f *is* C^2,
  (b) $\Gamma \in \text{GC}(1)$, *or*
  (c) f *is of class* $C^{1+\alpha+}$ *and* $\Gamma \in \text{GC}(1+\alpha)$, *for some* $\alpha \in (0, 1)$.

Theorem C is a strengthening of Theorem A because if $Df = \text{Id}$ on Γ, we may take g_Δ to be the identity map for all Δ. The proof makes use of the fact, from complex analysis, that a C^1 diffeomorphism with derivative pointwise everywhere in the conformal group is actually a conformal map. We refer the reader to [NV] for further details.

V. Concluding remarks

Theorems B and C are rigidity theorems in the mold of Denjoy's one-dimensional theorem. However, they both require additional hypotheses about the *values* of the derivative of f on Γ.

One goal is to try to establish these theorems for maps of Denjoy type, making use of the relation $hf = Rh$ and plane topology to understand Γ. Beyond that there is the problem of understanding the general homeomorphism semiconjugate to rigid rotation.

More important, however—and apparently more difficult—is the task of removing the extra derivative hypotheses, or else understanding the appropriate counterexamples. Here, it is the difficulty of making examples which is troublesome.

For example, one might try to construct a C^1 diffeomorphism of Sierpinski type having a single orbit of wandering *round* discs Δ_i (i.e., bounded by euclidean circles) with symmetry to the extent that $Df = (r_{i+1}/r_i)\,\mathrm{Id}$ on the boundary of Δ_i. However this is explicitly ruled out by Theorem $C(b)$. A C^1 example of Sierpinski type with $Df = \mathrm{Id}$ on Γ is not ruled out, but requires a dense orbit of fractal discs and this has resisted direct construction.

Work of Harrison [H] on related questions in the context of annulus diffeomorphisms has arrived at a similar stopping point. The field is ready for a new insight into smooth recurrence in two dimensions.

ADDED IN PROOF. About a year after this paper was written, P. McSwiggen [unpublished] tentatively constructed a nontrivial C^2 diffeomorphism of Denjoy type, thereby providing an affirmative answer to questions 1 and 3. The mapping is obtained as the holonomy of the one-dimensional unstable manifold foliation for a DA-type diffeomorphism of T^3. The method fails at C^3, suggesting the following revision:

QUESTION $1'$. Does there exist an example of a C^3 diffeomorphism of the 2-torus which is semiconjugate but not conjugate to R?

REFERENCES

[CFS] Cornfeld, Fomin, and Sinai, *Ergodic theory*, Springer-Verlag, New York, 1982.

[D] A. Denjoy, *Sur les courbes definies par les equations differentielles a la surface du tore*, J. Math. Pures Appl. (9) **11** (1932), 333–375.

[F] K. J. Falconer, *The geometry of fractal sets*, Cambridge Univ. Press, New York, 1985.

[Ha] G. R. Hall, *A generalization of Denjoy's theorem on diffeomorphisms of the circle*, Technical Summary Report No. 2516, Univ. of Wisconsin, Madison, Math. Research Center, May, 1983.

[H] J. Harrison, *Denjoy fractals*, Topology **28** (1989), 59–80.

[HN] J. Harrison and A. Norton, *Geometric integration on fractal curves in the plane*, Indiana Univ. Math. J., summer 1991, to appear.

[He] M. Herman, *Sur la conjugaison differentiable des diffeomorphisms du cercle a des rotations*, Inst. Hautes Études Sci. Publ. Math. **49** (1979), 5–233.

[HW] W. Hurewicz and H. Wallman, *Dimension theory*, Princeton Univ. Press, Princeton, N.J., 1941.

[KO] Y. Katznelson and D. Ornstein, *Diffeomorphisms of the circle*, preprint.

[M] R. Mackay, *A simple proof of Denjoy's theorem*, Nonlinearity (to appear).

[Ni] Z. Nitecki, *Differentiable dynamics*, M.I.T. Press, Cambridge, Mass., 1971.

[N1] A. Norton, *A critical set with nonnull image has large Hausdorff dimension*, Trans. Amer. Math. Soc. **296** (1986), 367–376.

[N2] ___, *Functions not constant on connected sets of critical points*, Proc. Amer. Math. Soc. **106** (1989), 397–405.

[N3] ___, *An area approach to wandering domains for smooth surface endomorphisms*, preprint.

[NV] A. Norton and J. Velling, *Denjoy-like theorems for diffeomorphisms of the 2-torus*, preprint.

[S] D. Sullivan, *Conformal dynamical systems*, Geometric Dynamics, Lecture Notes in Math., vol. 1007, Springer-Verlag, New York, 1983.

[W] H. Whitney, *A function not constant on a connected set of critical points*, Duke Math. J. **1** (1935), 514–517.

THE UNIVERSITY OF TEXAS AT AUSTIN, AUSTIN, TEXAS 78712

Contemporary Mathematics
Volume **117**, 1991

Rotations of Simply-Connected Regions and Circle-like Continua

JAMES T. ROGERS, JR.

1. Introduction

Let G be a simply-connected bounded domain in $\mathbf{C}$, and let $f: D \to G$ be a conformal, one-to-one mapping of the unit disk onto G. Let $R_\theta: z \to e^{i\theta} z$ be a rotation of D, and let $w_0 = f(0)$. The one-to-one, conformal self-mapping F of G defined by $F_\theta = f \circ R_\theta \circ f^{-1}$ is called an *intrinsic rotation of G about w_0*. The rotation F_θ is *extendible* if F_θ has a continuous extension to the closure of G. The extension of F_θ provides a "rotation" of ∂G.

A continuum is *circle-like* if it has finite open covers of arbitrarily small mesh whose nerves are circles. Many of the published examples of extendible intrinsic rotations have the property that ∂G is a circle-like continuum. On the other hand, continua theorists have been constructing their own brand of "rotation" of circle-like continua for many years.

We shall discuss some of these examples as well as the topological and prime end properties of ∂G that are implied by extendible intrinsic rotations. Particularly interesting is the case when ∂G is an indecomposable continuum. We shall compare these two notions of "rotation" in the case that ∂G is a circle-like continuum.

In §2, we abstract some material from [2] to motivate the study of extendible intrinsic rotations. For a complete exposition, see [2]. We also assume the reader is familiar with the theory of prime ends as expounded in [3].

1980 *Mathematics Subject Classification* (1985 *Revision*). Primary 30C35; Secondary 54F20.
This research was partially supported by NSF grant DMS-8600364.
This paper is in final form and no version of it will be submitted for publication elsewhere.

2. Motivation from complex analytic dynamics

Let $R: \overline{\mathbf{C}} \to \overline{\mathbf{C}}$ be a rational function defined on the Riemann sphere $\overline{\mathbf{C}} = \mathbf{C} \cup \{\infty\}$. If we write R as a quotient

$$R(z) = \frac{p(z)}{q(z)},$$

where $p(z)$ and $q(z)$ are polynomials with complex coefficients and no common factors, then the *degree* of R can be defined as

$$\deg(R) = \max\{\deg(p), \deg(q)\}.$$

We shall assume $\deg(R) \geq 2$.

A point z_0 is said to be *periodic* if $R^n(z_0) = z_0$, for some positive integer n. The smallest such integer n is called the *period* of z_0. The number $\lambda = (R^n)'(z_0)$ is called the *eigenvalue* of z_0.

A periodic point is said to be

(1) *superattracting* if $\lambda = 0$,
(2) *attracting* if $0 < |\lambda| < 1$,
(3) *neutral* if $|\lambda| = 1$, or
(4) *repelling* if $|\lambda| > 1$.

The *Julia set* $J(R)$ of R is the closure of the set of repelling periodic points of R. The *Fatou set* $F(R)$ is the complement of the Julia set.

The attracting and superattracting periodic points belong to the Fatou set, while the repelling periodic points belong, of course, to the Julia set. The behavior of the neutral periodic points is more interesting. We consider a fixed point z_0 (i.e., $R(z_0) = z_0$).

THEOREM 1. *A neutral fixed point z_0 belongs to the Fatou set $F(R)$ if and only if the Schröder Functional Equation has an analytic solution ϕ in some neighborhood U of z_0.*

The latter part of the theorem means that there exists an analytic homeomorphism $\phi: U \to D$ such that

$$\phi \circ R(z) = \lambda \phi(z).$$

If λ is a root of unity, then the Schröder Functional Equation does not have a solution, but if $\lambda = e^{2\pi i \alpha}$, where α is an irrational number poorly approximated by rationals, then it is a famous result of Siegel that there is a solution. Hence a neutral fixed point z_0 may belong to $F(R)$ or to $J(R)$.

Sullivan has classified the dynamics of periodic components of the Fatou set and divided them into five types. One type is a Siegel disk.

DEFINITION. Let G be a periodic component of $F(R)$ of period n. Let $S = R^n$. The domain G is a *Siegel disk* if G is simply connected and $S|G$ is analytically conjugate to a rotation.

An important question about Siegel disks remains undecided: Is every Siegel disk a Jordan domain? This question is a prime reason to study extendible intrinsic rotations.

3. Circle-like continua

A *continuum* is a compact, connected metric space. A continuum need not be locally connected, as the following example shows.

EXAMPLE 1. The topologist's $\sin 1/x$-curve is the union of two subsets A and B of the complex plane given by

$$A = \{(x, \sin 1/x): 0 < x \le 1/\pi\}, \qquad B = \{(0, y): -1 \le y \le 1\}.$$

A continuum is *chainable* or *arc-like* if it has finite open covers of arbitrarily small mesh whose nerves are arcs. The example above is chainable; a typical chain cover is indicated in Figure 1.

A continuum is *circle-like* if it has finite open covers of arbitrarily small mesh whose nerves are circles. One way to make a circle-like continuum is to double a chainable continuum by identifying two pairs of opposite endpoints. Figure 2 indicates a certain embedding of a doubled $\sin 1/x$-curve Y in the complex plane. This example is invariant under the rotation R_π of the plane through π radians.

Let G be a simply-connected domain in $\mathbf{C}$, $G \ne \mathbf{C}$, and let D be the unit disk in $\mathbf{C}$. Let $f: D \to G$ be a one-to-one, conformal map of D onto G, and let $f(0) = w_0$. Let $R_\theta: \mathbf{C} \to \mathbf{C}$ be the rotation through θ radians;

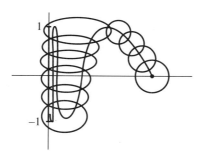

FIGURE 1

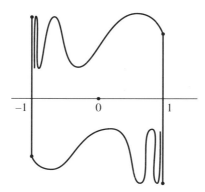

FIGURE 2

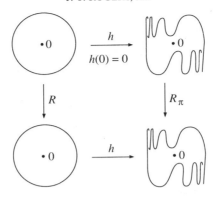

FIGURE 3

$R_\theta(z) = e^{i\theta} z$. The map $F_\theta = f \circ R_\theta \circ f^{-1}$ is called an *intrinsic rotation* of G about w_0. The map F_θ is said to be *extendible* if F_θ has a continuous extension to the closure of G in $\mathbf{C}$.

QUESTION. If F_θ is extendible, what can you say about the boundary of G?

A classical theorem of complex analysis asserts that f is extendible to a homeomorphism of $\overline{D}$ onto $\overline{G}$ if and only if G is a Jordan domain. It follows that if G is a Jordan domain, then F_θ is extendible. The converse does not hold, as the following example shows (see Figure 3).

EXAMPLE 2. Let Y be the doubled $\sin 1/x$-curve described in Figure 2. Let G be the bounded complementary domain of Y. Let $h: D \to G$ be a one-to-one, conformal map of D onto G satisfying $h(0) = 0$. The rotation R_π of G through π radians is an intrinsic rotation of G about 0, for the map $R = h^{-1} \circ R_\pi \circ h$ is a one-to-one conformal map of D onto D satisfying $R(0) = 0$ and $R'(0) = e^{i\pi}$. Hence $R = R_\pi$, by uniqueness.

One thing that continua theorists bring to the study of dynamical systems is a good supply of "pathological" examples, some of which occur quite naturally in dynamical systems. We shall illustrate this later in the paper, but we close this section with an example that will point out some limitations on the conversion of "topological rotations" into intrinsic rotations.

EXAMPLE 3. Consider the product of the standard Cantor set and the closed interval $[0, 1]$ as a subset of the plane. Add to this some horizontal line segments as indicated in Figure 4. The first such segment joins $(\frac{1}{3}, 0)$ and $(\frac{2}{3}, 0)$. The next two segments join $(\frac{1}{9}, 1)$ to $(\frac{2}{9}, 1)$ and $(\frac{7}{9}, 1)$ to $(\frac{8}{9}, 1)$, respectively. Continue to alternate the horizontal segments between the top and the bottom according to the length of the gaps of the Cantor set. Call this continuum A_1.

Shrink the horizontal line segments of A_1 to points to get another continuum A_2. This continuum is called the *Cajun accordion*. Both of these continua are chainable, and both of these continua admit a map onto the interval $[0, 1]$ whose fibers are precisely the arc components of the continuum

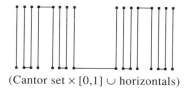

(Cantor set × [0,1] ∪ horizontals)

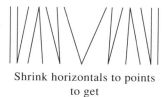

Shrink horizontals to points
to get
Cajun accordion

FIGURE 4

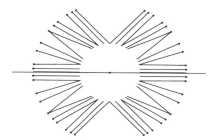

FIGURE 5. Zydeco accordion

(recall the standard map of the Cantor set onto the interval $[0, 1]$).

EXAMPLE 4. Double the Cajun accordion by identifying the two end arcs to get the zydeco accordion pictured in Figure 5 (the nomenclature of these examples is best appreciated after watching Zachary Richard at the Jazz Fest in New Orleans).

The zydeco accordion is a circle-like continuum used as a prototype for the construction of (topologically) homogeneous plane continua [15]. The decomposition into arc components yields a circle. A continua theorist views it as a continuum rich in rotations, but in the last section we shall see that it is meager with respect to extendible intrinsic rotations about the origin.

4. Regions with infinite families of extendible rotations

We adopt the following conventions throughout the rest of the paper. Let G be a simply-connected domain in $\mathbf{C}$, $G \neq \mathbf{C}$, and let D be the unit disk in $\mathbf{C}$. Let $f: D \to G$ be a fixed Riemann mapping function with $f(0) = w_0$.

Let

$$\Lambda = \{\theta \in \partial D : F_\theta \text{ is extendible}\}.$$

We shall assume during the rest of the paper that Λ is infinite. Note that if θ_1 and θ_2 belong to Λ, so does $\theta_1 \theta_2$. It follows that Λ is infinite if and only if Λ is dense in the boundary of D.

An important example of this occurs if $e^{2\pi i \alpha} \in \Lambda$, where α is an irrational number. In particular, to determine in this case when G is a Jordan domain is an analytic version of the Siegel disk problem for rational functions.

The following theorem combines some results of Burton Rodin, some classical results from continua theory, and a result of G. Schmieder to obtain ten necessary and sufficient conditions for G to be a Jordan domain.

THEOREM 2. *Suppose* $\Lambda = \{\theta \in \partial D : F_\theta \text{ is extendible}\}$ *is infinite. The following conditions are equivalent:*

(1) $\Lambda = \partial D$.
(2) Λ *is a* G_δ *subset of* ∂D.
(3) Λ *is a second category subset of* ∂D.
(4) *Each prime end of* G *is of the first kind.*
(5) *Every point of* ∂D *is accessible from* G.
(6) G *is a Jordan domain.*
(7) ∂D *is locally connected.*
(8) ∂D *is semi-locally connected.*
(9) ∂D *is aposyndetic.*
(10) ∂D *is freely decomposable.*
(11) *Some countable collection of intrinsic rotations is dense in the space of all intrinsic rotations with respect to uniform convergence on* G.

PROOF. The implications $(1) \Rightarrow (2)$ and $(2) \Rightarrow (3)$ are obvious. We say a word about $(3) \Rightarrow (4)$, since it is a slight improvement of Theorem 4 in [13]. Assuming the notation and results of that theorem, we know that $\mathscr{E} = \mathscr{E}_1 \cup \mathscr{E}_3$. Since $\mathscr{E}_2 = \varnothing$, $\mathscr{E}_1$ is residual in $\mathscr{E}$, and $\mathscr{E}_3$ is first category in $\mathscr{E}$ [3, p. 183]. If $P \in \partial D$ and $S(P) = \{e^{-i\theta}P : \theta \in \Lambda\}$, then $S(P)$ is second category in ∂D, by (3). Hence, $S(P)$ contains a point Q corresponding to a prime end Q' of the first kind. Some extendible rotation sends Q' onto P', the prime end corresponding to P. Since $\mathscr{E}_1 \Rightarrow \mathscr{E}_1$ under an extendible rotation, P corresponds to a prime end of the first kind.

Implications $(4) \Rightarrow (5)$, $(6) \Rightarrow (7)$, and $(6) \Rightarrow (1)$ are known. The implications $(5) \Rightarrow (6)$ and $(7) \Rightarrow (6)$ are due to Rodin ([13] and [11, Corollary 3]).

Properties (8), (9), and (10) are weak sorts of local connectivity; see F. B. Jones [7] for the definitions and proofs that they are equivalent to (6) in this situation.

Finally, G. Schmieder [17] has shown the equivalence of (11) and (6).

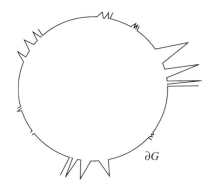

FIGURE 6. Moeckel example

5. Examples where ∂G is not a Jordan curve

There are three basic examples of domains G such that Λ is infinite and ∂G is not a Jordan curve. The first is due to R. Moeckel [10].

EXAMPLE 5. Let α be an irrational number. The Moeckel example is based on introducing a bad prime end (in this case, a prime end of Type 2) at a countable number of points of S^1 corresponding to the angles $\{e^{2\pi i\alpha} : n \in Z\}$. (See Figure 6.) The topological effect of this is to introduce limit bars of $\sin 1/x$-curves at this countable set of points. This example satisfies the following conditions:

(1) $\Lambda = \langle e^{2\pi i\alpha} \rangle$, α irrational.
(2) $\mathscr{E} = \mathscr{E}_1 \cup \mathscr{E}_2$.
(3) $\mathscr{E}_2$ is countably infinite.
(4) All impressions of prime ends are either points or radial line segments.
(5) The impressions of prime ends are pairwise disjoint.
(6) ∂G is a hereditarily decomposable circle-like continuum.

The Moeckel example can be viewed (topologically) as a continuum obtained from S^1 by blowing up a countable number of points into arcs. The zydeco accordion is a continuum obtained from S^1 by blowing up all points into arcs.

The next example, due to Pommerenke and Rodin [12], is entirely different.

EXAMPLE 6. Let α be an irrational number. The example of Pommerenke and Rodin has the following properties:

(1) $e^{2\pi i\alpha} \in \Lambda$.
(2) G is an unbounded domain that is star-like with respect to the origin.
(3) ∞ is an accessible point of ∂G.
(4) The intersection of the impressions of any two prime ends is $\{\infty\}$.

One can deduce several other properties about this example:

(5) ∂G is star-like with respect to ∞.

(6) Each impression is either a line segment or the point ∞.

(7) $\mathscr{E} = \mathscr{E}_1 \cup \mathscr{E}_2$.

(8) ∞ is an "explosion point" for the (connected) set of accessible points (the proof follows that of Mayer [9]).

Before we describe the third example, due to Herman [6] (see also Handel [5]), some definitions are in order.

A continuum is *indecomposable* if it is not the union of two of its proper subcontinua. A continuum is *hereditarily indecomposable* if each of its subcontinua is indecomposable. A *pseudocircle* is a hereditarily indecomposable, circle-like, separating planar continuum. R. H. Bing [1] described the pseudocircle as a candidate for a homogeneous plane continuum in 1951. It has lots of homeomorphisms, some extendible to the complex plane, and it seemed likely that with judicious rotations of the circular chain covers, any point could be moved to another point.

In 1968, however, L. Fearnley [4] and the author [16] independently showed that the pseudocircle is not homogeneous. The most extensive information on the homeomorphism group of the pseudocircle has been given by J. Kennedy and the author [8].

The pseudocircle is an example where the abundance of "topological" rotations carries over to an abundance of "analytic" rotations.

EXAMPLE 7. M. Herman [6] has constructed an example with the properties that

(1) ∂G is a pseudocircle, and

(2) $e^{2\pi i \alpha} \in \Lambda$, where α is a certain irrational number.

It follows from Theorem 3 of the next section that

(3) each impression $= \partial G$, and hence

(4) $\mathscr{E} = \mathscr{E}_2 \cup \mathscr{E}_3 \cup \mathscr{E}_4$,

(5) $\mathscr{E}_3$ is a dense G_δ subset, and

(6) $\mathscr{E}_2$ is a dense, first category subset of cardinality c.

6. Some new results

The Herman example shows that we must consider indecomposable continua in studying extendible intrinsic rotations. The following theorems of the author [14] should be helpful in this regard.

THEOREM 3. *If Λ is infinite, then ∂G is an indecomposable continuum if and only if each impression $= \partial G$.*

THEOREM 4. *If ∂G is indecomposable and Λ is infinite, then no composant is accessible at more than one point.*

Along a different line, notice that for a Jordan domain or the Moeckel example, each pair of impressions is disjoint, while in the examples of Herman and of Pommerenke and Rodin, each pair of impressions has a point in common. This leads to the following question:

QUESTION. If Λ is infinite, must it be the case that either each pair of impressions is disjoint or each pair of impressions has a point in common? In [14], the author has obtained the following result.

THEOREM 5. *The answer to the above question is yes in case* (a) ∂G *does not separate* $\overline{C}$, *or* (b) ∂G *is an irreducible separator of* $\overline{C}$.

A continuum X in $\overline{C}$ is an *irreducible separator* of $\overline{C}$ if X separates $\overline{C}$ but no proper subcontinuum separates. Circle-like, separating continua are examples of irreducible separators of $\overline{C}$; thus the examples of Herman and Moeckel have irreducible separators as boundaries.

In particular, the zydeco accordion is a circle-like separating continuum and hence an irreducible separator of $\overline{C}$. Thus we end with the following corollary.

COROLLARY 6. *If* Λ *is infinite, then* ∂G *is not* (*this embedding of*) *the zydeco accordion.*

REFERENCES

1. R. H. Bing, *Concerning hereditarily indecomposable continua*, Pacific J. Math. **1** (1951), 43–51.
2. P. Blanchard, *Complex analytic dynamics on the Riemann sphere*, Bull. Amer. Math. Soc. **11** (1984), 85–141.
3. E. F. Collingwood and A. J. Lohwater, *The theory of cluster sets*, Cambridge Univ. Press, New York, 1966.
4. L. Fearnley, *The pseudo-circle is not homogeneous*, Bull. Amer. Math. Soc. **75** (1969), 554–558.
5. M. Handel, *A pathological area preserving* C^∞ *diffeomorphism of the plane*, Proc. Amer. Math. Soc. **86** (1982), 163–168.
6. M. R. Herman, *Construction of some curious diffeomorphisms of the 2-sphere*, J. London Math. Soc. **34** (1986), 375–384.
7. F. B. Jones, *Aposyndetic continua and certain boundary problems*, Amer. J. Math. **63** (1941), 545–553.
8. J. Kennedy and J. Rogers, *Orbits of the pseudo-circle*, Trans. Amer. Math. Soc. **296** (1986), 327–340.
9. J. Mayer, *An explosion point for the set of endpoints of the Julia set of* $\lambda \exp(z)$, Ergodic Theory Dynamical Systems **10** (1990), 177–184.
10. R. Moeckel, *Rotations of the closures of some simply connected domains*, Complex Variables Theory Appl. **4** (1985), 223–232.
11. Ch. Pommerenke, *On the boundary continuity of conformal maps*, Pacific J. Math. **120** (1985), 423–430.
12. Ch. Pommerenke and B. Rodin, *Intrinsic rotations of simply connected regions. II*, Complex Variables Theory Appl. **4** (1985), 223–232.
13. B. Rodin, *Intrinsic rotations of simply connected regions*, Complex Variables Theory Appl. **2** (1984), 319–326.
14. J. T. Rogers, Jr., *Intrinsic rotations of simply connected regions and their boundaries*, Complex Variables Theory Appl. (to appear).

15. ____, *Almost everything you wanted to know about homogeneous, circle-like continua*, Topology Proc. **3** (1978), 169–174.

16. ____, *The pseudo-circle is not homogeneous*, Trans. Amer. Math. Soc. **148** (1970), 417–428.

17. G. Schmieder, *A characterization of Jordan domains in terms of conformal automorphisms*, Complex Variables Theory Appl. **2** (1984), 319–326.

DEPARTMENT OF MATHEMATICS, TULANE UNIVERSITY, NEW ORLEANS, LOUISIANA 70118

Contemporary Mathematics
Volume 117, 1991

Chaos: An Introduction to Some Topological Aspects

RICHARD M. SCHORI

1. Introduction

This paper is an introduction to some basic constructions that are proving useful in the study of chaotic dynamical systems. It is intended for students or nonexperts in the Theory of Continua. §2 contains some preliminary material on the Cantor Set and symbolic dynamics. §3 is on chaotic maps, §4 introduces inverse limit spaces with applications to symbolic dynamics, and §5 deals with inverse limits of chaotic maps and gives the fundamental result that the shift map on the inverse limit of chaotic maps is chaotic. Finally, in §6 we introduce the notion of attractor, give the appropriate inverse limit characterization, and conclude with an inverse limit construction of the solenoid attractor with the result that it is chaotic.

This paper is further designed for the reader who either wants an overview of the ideas or who wants to get involved with the details of the proofs. It is written partly from the perspective of a "theorem proving" class in that many of the easier proofs are omitted, whereas the discussion of the material is designed to help these proofs be accessible. Consequently, this paper could be the basis for a portion of a class or a seminar in which the students supply the proofs. Recommended references for inverse limit spaces are as follows: Basic properties can be found in the texts [5], [6], and [8]. Some tools and techniques can be found in [3] and [7]. Early applications of inverse limits to chaos are in [9] and [10], and samples of the recent work motivated by Joseph M. Martin and Marcy Barge are found in [1] and [2].

In these notes we adopt the philosophy of Robert L. Devaney as used in his book, *An Introduction to Chaotic Dynamical Systems* [4]. That is, we approach the topic of chaos *not* though systems of differential equations, but from the point of view of discrete dynamics, or roughly the study of continuous functions $f: X \to X$ from a space into itself. Although Devaney

---

1991 *Mathematics Subject Classification*. Primary 58F03, 58F13, 54B10, 54B99.
This paper is in final form and no version of it will be submitted for publication elsewhere.

only briefly mentions inverse limit spaces, I have tried to take a few of the main ideas from Devaney's book and express them in the language and format of inverse limits. This paper evolved out of a set of notes developed for a beginning graduate level topology seminar at Oregon State University. The author thanks Robert M. Burton and Marcy Barge for helpful conversations.

2. The Cantor set and sequence space

The student of chaos soon realizes that the celebrated Cantor set is inherent in and fundamental to the concept of chaos. Consequently, we start by introducing the Cantor set. If not the most important example in topology, it is certainly one of the most important. Furthermore it is a fractal, which means, from an intuitive point of view, that it is self-similar under magnification.

The *Cantor set* or standard middle third Cantor set is defined as follows: Let A_0 be the closed unit interval $[0, 1]$, let $A_1 = [0, \frac{1}{3}] \cup [\frac{2}{3}, 1]$, and in general, A_{n+1} is obtained from A_n by removing the open middle third interval from each interval in A_n. Then the Cantor set $C = \bigcap_{n=0}^{\infty} A_n$. It is convenient to label the subintervals of the A_n as follows: $I_0 = [0, \frac{1}{3}]$, $I_1 = [\frac{2}{3}, 1]$, $I_{00} = [0, \frac{1}{9}]$, $I_{01} = [\frac{2}{9}, \frac{1}{3}]$, $I_{10} = [\frac{2}{3}, \frac{7}{9}]$, $I_{11} = [\frac{8}{9}, 1]$, etc. Thus $A_1 = I_0 \cup I_1$, $A_2 = I_{00} \cup I_{01} \cup I_{10} \cup I_{11}$, etc. A natural presentation of the Cantor set is the base-3 representation of $[0, 1]$ with sequences of 0's and 2's.

LEMMA 1. *The Cantor set* $C = \{.b_1 b_2 b_3 \cdots : $ *this is the base-3 representation of real numbers where each* $b_i \in \{0, 2\}\}$.

Lemma 1 will be useful a little later but we introduce it now to help motivate sequence space which is the starting point for *symbolic dynamics*, an important tool in describing examples of chaotic behavior. We shall see that the space Σ defined next is homeomorphic to the Cantor set and hence represents another way of expressing the Cantor set. Its introduction will pay many dividends.

Let Σ denote *sequence space* where $\Sigma = \{(a_1, a_2, \ldots) : a_i \in \{0, 1\}\}$ with a metric defined on Σ by $d(\mathbf{a}, \mathbf{b}) = 1/k$, if $k = \min\{n : a_n \neq b_n\}$. Thus, two members of Σ are distance $1/k$ apart if they agree in their first $(k-1)$ coordinates and disagree in their kth coordinate.

To better understand Σ we will first investigate some basic properties.

EXERCISE 2. Verify that d is a metric for Σ.

Another metric for Σ can be defined as follows: If $\mathbf{a} = (a_1, a_2, \ldots)$ and $\mathbf{b} = (b_1, b_2, \ldots)$, define the distance between them by

$$d'(\mathbf{a}, \mathbf{b}) = \sum_{i=1}^{\infty} \frac{|a_i - b_i|}{2^i}.$$

EXERCISE 3. Verify that d' is a metric for Σ and that d and d' are equivalent metrics.

Define $f: \Sigma \rightarrow C$ by $f(a_1, a_2, \ldots) = \sum_{n=1}^{\infty} 2a_n/3^n$. If $b_n = 2a_n$, then $f(a_1, a_2, \ldots) = .b_1, b_2 \cdots$ in its base-3 representation.

LEMMA 4. *For the map $f: \Sigma \rightarrow C$ defined above, we have*

$$f(a_1, a_2, \ldots) = \bigcap_{n=1}^{\infty} I_{a_1 a_2 \cdots a_n}.$$

(Note that $I_{a_1} \supset I_{a_1 a_2} \supset I_{a_1 a_2 a_3} \supset \cdots$ and the lengths of these intervals are going to 0 and hence the intersection of this nested sequence is a single point.)

THEOREM 5. *The map $f: \Sigma \rightarrow C$ is a homeomorphism $(\cong)$.*

Thus we see that sequence space Σ is topologically the Cantor set. The standard topological characterization of the Cantor Set [6] says that any compact totally disconnected perfect metric space is homeomorphic to the Cantor set.

3. Chaotic maps

We give a definition of a *chaotic* map and then verify that Σ with the shift map is chaotic. If $f: X \rightarrow X$ is a map and $x \in X$, then x is a *periodic point* of f if there exists a positive integer n such that $f^n(x) = x$, and the *orbit* of x, $O_x = \{f^n(x): n = 0, 1, 2, \ldots\}$. A set $A \subset X$ is *dense* in X if (the closure of A) $\overline{A} = X$. We will use the quantifiers $\forall, \exists$ to mean "for each," "there exists," respectively.

DEFINITION. Let X be a compact metric space. Then a mapping $f: X \rightarrow X$ is said to be *chaotic* if

1. The periodic points of f are dense in X.

2. Map f has *sensitive dependence on initial conditions* (SIC). That is, there is a distance $\delta > 0$, (thought of as large) such that for each $x \in X$, there exists $y \in X$, arbitrarily close to x, such that for some integer n, the distance from $f^n(x)$ to $f^n(y)$ is greater than δ. With quantifiers, f has SIC if $\exists \delta > 0 \ \forall x \in X \ \forall \varepsilon > 0 \ \exists y \in X$ such that $d(x, y) < \varepsilon$ and $\exists n$ such that $d(f^n(x), f^n(y)) \geq \delta$.

3. Map f is *topologically transitive*, that is, for each pair on nonempty open sets U, V in X, there exists an $n \geq 0$ such that $f^n(U) \cap V \neq \varnothing$.

Alternately,

$3'$. There exists a point $x \in X$ with a dense orbit O_x.

We also say that the pair (X, f) is *chaotic*.

COMMENTS. a. Condition 2 has become the standard accepted notion of chaos, that is, a deterministic system (in this case $f: X \rightarrow X$) with the property that you can have two points x, y that are arbitrarily close (perhaps closer than the round off error on your computer) such that their orbits diverge dramatically. A popular scenario is the "butterfly effect," that is, a butterfly flapping its wings in China can cause a hurricane in New York.

b. It is easy to prove that condition $3'$ always implies condition 3, and that if X is a complete metric space, a proof using the Baire Category Theorem can be constructed to prove that condition 3 implies condition $3'$.

LEMMA 7. *For a map $f: X \to X$, if there exists a dense orbit in X, then f is topologically transitive.*

Henceforth, we will use whichever of 3 or $3'$ is more convenient.

Shift map. We are now ready to define a map σ on Σ and prove that it is chaotic. Let the *shift* map $\sigma: \Sigma \to \Sigma$ be defined by

$$\sigma(a_1, a_2, \ldots) = (a_2, a_3, \ldots).$$

This map is clearly two-to-one since the a_1 information is lost and a_1 has two possible values, namely 0 and 1. By using the metric d on Σ, two sequences in Σ are within $1/k$ of each other if they agree on their first $k-1$ coordinates. This, with a careful selection of points, yields the following theorem.

THEOREM 8. *The map σ is chaotic.*

PROOF. (Property 1) Let $\mathbf{a} = (a_1, a_2, \ldots)$ be an arbitrary point of Σ and let $\varepsilon > 0$. We need to find a periodic point $\mathbf{b}$ in Σ such that $d(\mathbf{a}, \mathbf{b}) < \varepsilon$. Let k be a positive integer such that $1/k < \varepsilon$, and let $\mathbf{b} = (a_1, \ldots, a_k, a_1, \ldots, a_k, \ldots)$ be the point where the first k coordinates of $\mathbf{a}$ are repeated ad infinitum. Then $\sigma^k(\mathbf{b}) = \mathbf{b}$ and $\mathbf{b}$ is sufficiently close to $\mathbf{a}$.

(Property 2) Let $\delta = 1$, let $\mathbf{a} = (a_1, a_2, \ldots)$ be an arbitrary point of Σ, let $\varepsilon > 0$, and let $1/k < \varepsilon$. Then if $\mathbf{b}$ is the point of Σ that agrees with $\mathbf{a}$ for the first $k - 1$ coordinates and differs from $\mathbf{a}$ in the kth coordinate, then $d(\mathbf{a}, \mathbf{b}) < \varepsilon$ and $d(\sigma^k(\mathbf{a}), \sigma^k(\mathbf{b})) = 1$.

(Property 3) Let $\mathbf{a}$ be the point of Σ whose coordinates are successively all possible one-bit words $(0, 1)$, and then all possible two-bit words $(0,0, \; 0,1, \; 1,0, \; 1,1)$, and so forth. Then $\mathbf{a}$ has a dense orbit. □

We are now ready to show some further applications of Σ and σ.

EXERCISE 9. The map $\beta: \Sigma \to [0, 1]$ defined by $\beta(a_1, a_2, \ldots) = .a_1 a_2 a_3 \cdots$ in binary representation is a continuous and onto map. By studying Lemma 1 and Theorem 5 show that $\beta f^{-1}: C \to [0, 1]$ is two-to-one on the endpoints of the complementary intervals of the Cantor set and in fact takes the endpoints of $(\frac{1}{3}, \frac{2}{3})$ to $\frac{1}{2}$, the endpoints of $(\frac{1}{9}, \frac{2}{9})$ to $\frac{1}{4}$, the endpoints of $(\frac{7}{9}, \frac{8}{9})$ to $\frac{3}{4}$, etc.

We now give the circle S^1 a convenient parameterization. Let $S^1 = \mathbb{R} \bmod 1$, i.e., for $x, y \in \mathbb{R}$, $x \equiv y$ iff $x - y = n$, for some $n \in \mathbb{Z}$. We have parameterized the circle with the interval $[0, 1]$ where 0 has been identified with 1.

Define $g: S^1 \to S^1$ by $g(x) = 2x$ and define $\alpha: \Sigma \to S^1$ by

$$\alpha(a_1, a_2, \ldots) = \sum_{n=1}^{\infty} \frac{a_n}{2^n} = .a_1 a_2 a_3 \cdots$$

in binary representation.

EXERCISE 10. The map α is continuous.

LEMMA 11. *The diagram commutes, that is, $\alpha \sigma = g \alpha$.*

$$
\begin{array}{ccc}
\Sigma & \xrightarrow{\sigma} & \Sigma \\
\downarrow \alpha & & \downarrow \alpha \\
S^1 & \xrightarrow{g} & S^1
\end{array}
$$

Combining the proof that (Σ, σ) is chaotic and the above lemma it is easy to prove the following theorem.

THEOREM 12. *The pair (S^1, g) is chaotic.*

We will now look at the general situation of developing tools for determining when a given $g: Y \to Y$ is chaotic.

DEFINITION. Let X and Y be topological spaces and let $f: X \to X$ and $g: Y \to Y$ be maps. A surjective map $\alpha: X \to Y$ is said to be a *semi-conjugacy* if $\alpha f = g \alpha$, that is, the following diagram commutes:

$$
\begin{array}{ccc}
X & \xrightarrow{f} & X \\
\downarrow \alpha & & \downarrow \alpha \\
Y & \xrightarrow{g} & Y
\end{array}
$$

In this case f and g are called *semi-conjugate*. If, in addition, α is a homeomorphism of X onto Y, then α is called a *topological conjugacy*, and f and g are said to be *topologically conjugate*. Note that $\alpha f = g \alpha$ implies $\alpha f^n = g^n \alpha$. In fact $\alpha f^2 = (g \alpha) f = g(\alpha f) = g(g \alpha) = g^2 \alpha$, etc. The image of an f orbit by a semi-conjugacy is a g orbit, while a topological conjugacy sends f orbits to g orbits and preserves their topological properties.

For the following two lemmas assume that the next diagram exists and commutes.

$$
\begin{array}{ccc}
X & \xrightarrow{f} & X \\
\downarrow \alpha & & \downarrow \alpha \\
Y & \xrightarrow{g} & Y
\end{array}
$$

LEMMA 13. *If (X, f) is chaotic, and α is a topological conjugacy, then (Y, g) is also chaotic.*

LEMMA 14. (a) *If the periodic points of f are dense in X and α is a semi-conjugate, then the periodic points of g are dense in Y.*

(b) *If there exists a dense orbit O_x in X, and α is a semi-conjugate, then there exists a dense orbit O_y in Y.*

EXERCISE 15. By letting Y be a singleton, verify that if (X, f) is chaotic and α is a semi-conjugate, then (Y, g) need not have SIC.

PROBLEM 16. Under what conditions is it true that if $f: X \to X$ has SIC and α is a semi-conjugate, that $g: Y \to Y$ has SIC? Try Y compact and α is at most n-to-one for some positive integer n. (In most applications, such as in the proof of Theorem 12, the SIC property of f clearly illustrates how to prove the SIC property of g.)

4. Inverse limit spaces

Inverse limit spaces yield powerful techniques for constructing complicated spaces and maps from simple spaces and maps. Two examples from topology that motivated me in the subject years ago have resurfaced as important examples in chaos. The first is the limit of the inverse sequence

$$S^1 \xleftarrow{g} S^1 \xleftarrow{g} I \xleftarrow{g} \cdots$$

where each *coordinate space* is a circle and each map is the $\times 2$ map, $g(x) = 2x$. The limit of this sequence is what is known in topology as the *dyadic solenoid* and in chaos it is known as the *solenoid attractor*. The second example is the limit of the inverse sequence

$$I \xleftarrow{f} I \xleftarrow{f} I \xleftarrow{f} \cdots$$

where each coordinate space $I = [0, 1]$ and each bonding map $f: I \to I$ is the tent map

$$f(x) = \begin{cases} 2x, & 0 \leq x \leq \frac{1}{2}, \\ 2 - 2x, & \frac{1}{2} \leq x \leq 1. \end{cases}$$

The limit of the inverse sequence has been known among topologists as either Knaster's indecomposable continuum or the "bucket handle." In more recent years (with appropriate discussions of the dynamics involved) it has been known as Smale's horseshoe. We will now make these ideas precise.

DEFINITION. An *inverse sequence* is a double sequence $(X_n, f_n)_{n=1}^{\infty}$ such that each *coordinate space* is a topological space and each *bonding map* $f_n: X_{n+1} \to X_n$ is continuous

$$X_1 \xleftarrow{f_1} X_2 \xleftarrow{f_2} X_3 \cdots.$$

The *limit* of the inverse sequence is the set

$$X_\infty = \varprojlim(X_n, f_n)$$
$$= \left\{ (x_n) \in \prod_{n=1}^{\infty} X_n : \forall n \geq 1, \ f_n(x_{n+1}) = x_n \right\}$$

topologized with the relativized product topology. We will expand on that below but first some intuition. The inverse limit is a very special subset of the product $\prod_{n=1}^{\infty} X_n$, namely, just those points (x_n) whose coordinates map down to the next lowest one, i.e., $f(x_{n+1}) = x_n$. This set can be thought of as a diagonal or graph in the product. Note that if the sequence consisted of just one function f_1, then the "inverse limit" would be the set $\{(x_1, x_2): f_1(x_2) = x_1\} \subset X_1 \times X_2$ which is precisely the graph of f_1 and is homeomorphic to

X_2. Likewise, $\{(x_1, x_2, \ldots, x_n): \text{ for each } 1 \leq i < n, \ f(x_{i+1}) = x_i\}$ is homeomorphic to X_n, and, correspondingly, X_∞ can be thought of as a set-wise limit in $\prod_{n=1}^{\infty} X_n$ of copies of X_n, as $n \to \infty$.

Let π_k denote the *natural projection* from both $\prod_{n=1}^{\infty} X_n$ and its subset X_∞ onto X_k defined by $\pi_k((x_n)) = x_k$.

A *basis* β for a topology τ is a subcollection of τ such that if $U \in \tau$ and $p \in U$, then there exists $B \in \beta$ such that $p \in B \subset U$. The *product topology* on $\prod_{n=1}^{\infty} X_n$ is typically defined in terms of a basis which consists of all sets of the form

$$\{\pi_{n_1}^{-1}(U_{n_1}) \cap \cdots \cap \pi_{n_k}^{-1}(U_{n_k}): k \geq 1, \text{ and each } U_{n_i} \text{ is open in } X_{n_i}\}.$$

An important feature of inverse limits is that when we restrict the topology of the entire product space down to its subspace X_∞, we obtain a significant simplification.

LEMMA 17. *The collection* $\{\pi_k^{-1}(U_k): k \geq 1 \text{ and } U_k \text{ is open in } X_k\}$ *is a basis for the topology on* X_∞.

The next two propositions are to give the reader some technical understanding of and intuitive feel for inverse limits.

PROPOSITION 18. *If* $X_1 \supset X_2 \supset \cdots$ *and the bonding maps are injections, then we can think of the inverse limit as the intersection of the spaces.*

$$X_1 \xleftarrow{i_1} X_2 \xleftarrow{i_2} X_3 \xleftarrow{i_3} \cdots \bigcap_{n=1}^{\infty} X_n$$

Explicitly, if each $i_n: X_{n+1} \to X_n$ *is the injection map and* $X_\infty = \varprojlim(X_n, i_n)$, *then* $h: X_\infty \to \bigcap_{n=1}^{\infty} X_n$ *defined by* $h(x, x, x, \ldots) = x$ *is a homeomorphism.*

EXERCISE 19. Let A_n be as in the definition of the Cantor set and let $i_n: A_{n+1} \to A_n$ be the injection map. Then $\varprojlim(A_n, i_n) \cong C$.

PROPOSITION 20. *In an inverse sequence, if all the coordinate spaces are homeomorphic to a given space* X *and each bonding map is a homeomorphism, then the inverse limit is also homeomorphic to* X. *Explicitly, in the inverse sequence*

$$X_1 \xleftarrow{h_1} X_2 \xleftarrow{h_2} X_3 \xleftarrow{h_3} \cdots$$

if each $X_n \cong X_1$ *and each* h_n *is a homeomorphism, then* $h: \varprojlim(X_n, h_n) \to X_1$ *defined by* $h(x_1, x_2, \ldots) = x_1$ *is a homeomorphism.*

Induced map between inverse limits.

EXERCISE 2.1. Let (X_n, f_n) and (Y_n, g_n) be inverse sequences, and for each $n \geq 1$ let $h_n: X_n \to Y_n$ be a function such that $h_n f_n = g_n h_{n+1}$. Then there is an induced function $h_\infty: X_\infty \to Y_\infty$ defined by $h_\infty((x_n)) = (h_n(x_n))$.

Furthermore, if each $h_n: X_n \to Y_n$ is continuous, then $h_\infty: X_\infty \to Y_\infty$ is continuous.

$$
\begin{array}{ccccccccc}
X_1 & \xleftarrow{f_1} & X_2 & \xleftarrow{f_2} & X_3 & \xleftarrow{f_3} & \cdots & & X_\infty \\
\downarrow h_1 & & \downarrow h_2 & & \downarrow h_3 & & \cdots & & \downarrow h_\infty \\
Y_1 & \xleftarrow{g_1} & Y_2 & \xleftarrow{g_2} & Y_3 & \xleftarrow{g_3} & \cdots & & Y_\infty
\end{array}
$$

We will now illustrate this with an inverse limit presentation of the Cantor set. Let $S_0 = \{s\}$, $S_1 = \{s_0, s_1\}$, $S_2 = \{s_{ij} : i, j = 0, 1\}$, and in general, $S_n = \{s_{i_1 i_2 \cdots i_n} : i_1, i_2, \ldots, i_n = 0, 1\}$. Thus S_n consists of 2^n points. Endow S_n with the discrete topology. Define $f_n: S_{n+1} \to S_n$ by $f_n(s_{i_1 i_2 \cdots i_{n+1}}) = s_{i_1 i_2 \cdots i_n}$.

EXERCISE 22. The inverse limit, $\varprojlim (S_n, f_n) \cong C$.

Recall that $A_n = \bigcup I_{i_1 i_2 \cdots i_n}$, where $i_j \in \{0, 1\}$. Then A_n consists of the union of 2^n subintervals. Define $r_n: A_n \to S_n$ by $r_n(I_{i_1 i_2 \cdots i_n}) = s_{i_1 i_2 \cdots i_n}$. Thus r_n collapses each interval in A_n to a point, the corresponding point in S_n.

$$
\begin{array}{ccccccccc}
A_1 & \xleftarrow{i_1} & A_2 & \xleftarrow{i_2} & A_3 & \xleftarrow{i_3} & \cdots & & C \\
\downarrow r_1 & & \downarrow r_2 & & \downarrow r_3 & & \cdots & & \downarrow r_\infty \\
S_1 & \xleftarrow{f_1} & S_2 & \xleftarrow{f_2} & S_3 & \xleftarrow{f_3} & \cdots & & C
\end{array}
$$

EXERCISE 23. The induced map $r_\infty: C \to C$ is a homeomorphism. (Note that each r_n is not one-to-one, but r_∞ is one-to-one.)

A product space version of the Cantor set. If X is a topological space, then by X^ω we mean the countable infinite product of X with the product topology. (Let $\omega = \{1, 2, \ldots\}$.) It is well understood that we can identify X^ω with the set of all functions from ω into X which in turn is the set of all sequences of points in X. If $X = \{0, 1\}$, then $X^\omega = \Sigma$ and the product topology on X^ω coincides with the topology induced by the metric $d(\mathbf{a}, \mathbf{b}) = 1/k$, if $k = \min\{n: a_n \neq b_n\}$.

EXERCISE 24. If $\{0, 1\}$ is given the discrete topology, then $C \cong \{0, 1\}^\omega$. It readily follows that $C \times C \cong C$.

5. Chaotic maps and inverse limits

We will first show that double sequence space is the inverse limit of sequence space with the shift map as bonding map.

Define the *double sequence* space Σ' by

$$\Sigma' = \{(\ldots, a_{-2}, a_{-1}, a_0, a_1, a_2, \ldots) : a_i \in \{0, 1\}\}$$

with metric $d(\mathbf{a}, \mathbf{b}) = 1/k$, if $k = \min\{|n| : a_n \neq b_n, n \in \mathbb{Z}\}$ and $k \neq 0$, and $d(\mathbf{a}, \mathbf{b}) = 2$ if $a_0 \neq b_0$. (You can think of Σ' as the product of two Cantor sets, which of course is a Cantor set.)

LEMMA 25. *The map $g: \Sigma \to \Sigma'$ defined by $g(a_1, a_2, \ldots) = (\ldots, a_4, a_2, \overset{\circ}{a}_1, a_3, a_5, \ldots)$ is a homeomorphism. Also $\Sigma \times \Sigma \cong \Sigma'$ and thus $\Sigma \times \Sigma \cong \Sigma$. (These properties should not be surprising since $\Sigma \cong C$.)*

We will now show that it is natural to think of Σ' as an inverse limit of copies of Σ where each bonding map is the shift map σ.

$$\Sigma \xleftarrow{\sigma} \Sigma \xleftarrow{\sigma} \Sigma \xleftarrow{\sigma} \cdots \Sigma'$$

LEMMA 26. *The double sequence space* $\Sigma' \cong \varprojlim (\Sigma, \sigma)$, *where*

$$\sigma(a_0, a_1, a_2, \ldots) = (a_1, a_2, \ldots)$$

is the shift map. In particular, $h: \varprojlim (\Sigma, \sigma) \to \Sigma'$ *defined by*

$$h((a_1, a_2, \ldots), (a_0, a_1, a_2, \ldots), (a_{-1}, a_0 a_1, \ldots), \ldots)$$
$$= (\ldots, a_{-2}, a_{-1}, a_0 a_1, a_2, \ldots)$$

is a homeomorphism.

The next lemma shows that the shift map $\sigma: \Sigma \to \Sigma$ naturally induces a shift map $\tilde{\sigma}$ on Σ'. We shall see that $\tilde{\sigma}$ will be a homeomorphism whereas σ is 2-1 and $\tilde{\sigma}$ inherits all the dynamics of σ.

LEMMA 27. *The diagram*

$$
\begin{array}{ccccccccc}
\Sigma & \xleftarrow{\sigma} & \Sigma & \xleftarrow{\sigma} & \Sigma & \xleftarrow{\sigma} & \cdots & \Sigma' & (\ldots, a_{-2}, a_{-1}, \overset{\circ}{a}_0, a_1, \ldots) \\
\downarrow \sigma & & \downarrow \sigma & & \downarrow \sigma & & \cdots \downarrow \tilde{\sigma} & & \downarrow \tilde{\sigma} \\
\Sigma & \xleftarrow{\sigma} & \Sigma & \xleftarrow{\sigma} & \Sigma & \xleftarrow{\sigma} & \cdots & \Sigma' & (\ldots, a_{-1}, a_0, \overset{\circ}{a}_1, a_2, \ldots)
\end{array}
$$

induces a shift map $\tilde{\sigma}: \Sigma' \to \Sigma'$ *described at the right of the diagram and* $\tilde{\sigma}$ *is a homeomorphism. The proof that* (Σ, σ) *is chaotic adapts naturally to the following theorem.*

THEOREM 28. *The pair* $(\Sigma', \tilde{\sigma})$ *is chaotic.*

We have the corresponding result for the solenoid.

LEMMA 29. *If* $\mathbb{S} = \varprojlim (S^1, g)$, *where* $g(t) = 2t$, *then the diagram*

$$
\begin{array}{ccccccccc}
S^1 & \xleftarrow{g} & S^1 & \xleftarrow{g} & S^1 & \xleftarrow{g} & \cdots & \mathbb{S} & (t_1, t_2, t_3, \ldots) \\
\downarrow g & & \downarrow g & & \downarrow g & & \downarrow \tilde{g} & & \downarrow \tilde{g} \\
S^1 & \xleftarrow{g} & S^1 & \xleftarrow{g} & S^1 & \xleftarrow{g} & \cdots & \mathbb{S} & (g(t_1), t_1, t_2, \ldots)
\end{array}
$$

induces a shift map $\tilde{g}$ *on* $\mathbb{S}$ *described at the right of the diagram and* $\tilde{g}$ *is a homeomorphism. (Recall that each* $t_i \in \mathbb{R}(\mathrm{mod}\, 1)$ *and thus if* $t_1 = .a_1 a_2 a_3 \cdots$ *in base 2, then* $g(t_1) = .a_2 a_3 \ldots .$)

THEOREM 30. *The pair* $(\mathbb{S}, \tilde{g})$ *is chaotic.*

EXERCISE 31. Verify that the map $\tilde{\alpha}: \Sigma' \to \mathbb{S}$ defined by

$$\tilde{\alpha}(\ldots, a_{-1}, a_0, a_1, a_2, \ldots) = (.a_1 a_2 a_3 \ldots, .a_0 a_1 a_2, \ldots, .a_{-1} a_0 a_1 \ldots, \ldots)$$

is a semi-conjugate of the maps $\tilde{\sigma}: \Sigma' \to \Sigma'$ and $\tilde{g}: \mathbb{S} \to \mathbb{S}$.

We are now well motivated to look at the general case. If X is a metric space and $f: X \to X$ is continuous and onto, let $\mathbb{X} = \mathbb{X}(f)$ denote the limit of the following inverse sequence and let $\tilde{f}: \mathbb{X} \to \mathbb{X}$ be the map induced by the following diagram.

$$
\begin{array}{ccccccccc}
X & \xleftarrow{f} & X & \xleftarrow{f} & X & \xleftarrow{f} & \cdots & & X \\
\downarrow f & & \downarrow f & & \downarrow f & & & & \downarrow \tilde{f} \\
X & \xleftarrow{f} & X & \xleftarrow{f} & X & \xleftarrow{f} & \cdots & & X
\end{array}
$$

We need a specific metric on $\mathbb{X}$, so let $\mathbb{X} = \{(x_1, x_2, \ldots) : x_n \in X$ and if $n = 1, 2, \ldots$, then $f(x_{n+1}) = x_n\}$ with metric

$$
D((x_1, x_2, \ldots), (y_1, y_2, \ldots)) = \sum_{i=1}^{\infty} \frac{d(x_i, y_i)}{2^i},
$$

where d denotes the metric on X. This is the usual product metric and thus it induces the inverse limit topology on $\mathbb{X}$ as desired. Also assume that diameter$(X) \leq 1$.

We are now ready to work toward the result that if (X, f) is chaotic, then so is $(\mathbb{X}, \tilde{f})$. First we need a few lemmas.

LEMMA 32. *For each $\underline{x} \in \mathbb{X}$ and $\varepsilon > 0$, there exists a positive integer k and an $\alpha > 0$ such that if $\underline{y} \in \mathbb{X}$, where $d(y_k, x_k) < \alpha$, then $D(\underline{x}, \underline{y}) < \varepsilon$.*

PROOF. If $\underline{x} \in \mathbb{X}$ and $\varepsilon > 0$ are given, let k be such that $2^{-k} < \varepsilon/2$. Using the continuity of the bonding maps, for each $i = 1, \ldots, k - 1$, there exists $\alpha_i > 0$ such that if $y_k \in X_k$, where $d(y_k, x_k) < \alpha_i$, then $d(f^i(y_k), f^i(x_k)) < \varepsilon/2$. Note that $f^i(y_k) = y_{k-i}$ and $f^i(x_k) = x_{k-i}$. Let $\alpha = \min\{\varepsilon/2, \alpha_1, \ldots, \alpha_{k-1}\}$. Now, if $\underline{y} \in \mathbb{X}$ where $d(y_k, x_k) < \alpha$, then for each $i = 1, \ldots, k$, $d(x_i, y_i) < \varepsilon/2$ and

$$
D(\underline{x}, \underline{y}) = \sum_{i=1}^{\infty} \frac{d(x_i, y_i)}{2^i} \leq \frac{\varepsilon}{2}\left(\sum_{i=1}^{k} \frac{1}{2^i}\right) + \sum_{i=k+1}^{\infty} \frac{1}{2^i}
$$

$$
\leq \frac{\varepsilon}{2} + 2^{-k} < \varepsilon. \quad \square
$$

LEMMA 33. *If X is a compact metric space, and $f: X \to X$ is SIC, then $\tilde{f}: \mathbb{X} \to \mathbb{X}$ is SIC.*

PROOF. Let $\delta > 0$ be given from the assumption that $f: X \to X$ is SIC. Let $\underline{x} \in \mathbb{X}$ and $\varepsilon > 0$. Assume that $\varepsilon < \delta$. Apply the previous lemma to obtain k and $\alpha > 0$ such that if $\underline{y} \in \mathbb{X}$, where $d(x_k, y_k) < \alpha$, then $D(\underline{x}, \underline{y}) < \varepsilon$. Take such a $\underline{y}$. Since f is SIC, there exists a positive integer m such that $d(f^m(x_k), \overline{f^m(y_k)}) = d(x_{k-m}, y_{k-m}) \geq \delta$. Since $d(x_i, y_i) < \varepsilon/2 < \delta$, for $i = 1, \ldots, k$, we have $m > k$. Now, if $\underline{x} = (x_1, x_2, \ldots) \in \mathbb{X}$, then $\tilde{f}(\underline{x}) = (x_0, x_1, x_2, \ldots)$, and in general $\tilde{f}^{i+1}(\underline{x}) = (x_{-i}, \ldots, x_0, x_1, \ldots)$. Thus, $\tilde{f}^{m-k+1}(\underline{x}) = (x_{k-m}, \ldots, x_0, x_1, \ldots)$ and hence, if $n = m - k + 1$, then $D(\tilde{f}^n(\underline{x}), \tilde{f}^n(\underline{y})) \geq \delta/2$. Consequently, $\tilde{f}$ is SIC. $\square$

LEMMA 34. *If $\tilde{f}:\mathbb{X} \to \mathbb{X}$ is induced by $f:X \to X$, then for $U \subset X_k = X$ we have*

(a) $\tilde{f}^n(\pi_k^{-1}(U)) = \pi_k^{-1}(f^n(U))$, *for $n = 1, 2, \ldots$, and*

(b) *if $k < r$ and $U_k \subset X_k$, then $\pi_k^{-1}(U_k) = \pi_r^{-1}(U_r)$, where $U_r = (f^{r-k})^{-1}(U_k) \subset X_r$.*

LEMMA 35. *If $f:X \to X$ is topologically transitive, then $\tilde{f}:\mathbb{X} \to \mathbb{X}$ is also topologically transitive.*

PROOF. Let $\mathbb{U}, \mathbb{V}$ be open sets in $\mathbb{X}$. We can assume that they are basic open sets $\pi_m^{-1}(U_m)$ and $\pi_k^{-1}(V_k)$, respectively. If $m < k$, let $U_k = (f^{k-m})^{-1}(U_m)$. Then $\pi_k^{-1}(U_k) = \pi_m^{-1}(U_m)$, and if $k < m$, do the analogous thing. Assuming f is topologically transitive, there exists $n \geq 1$ such that $f^n(U_k) \cap V_k \neq \varnothing$. The conclusion we want, $\tilde{f}^n(\mathbb{U}) \cap \mathbb{V} \neq \varnothing$, now follows since

$$\tilde{f}^n(\mathbb{U}) \cap \mathbb{V} = \tilde{f}^n(\pi_k^{-1}(U_k)) \cap \mathbb{V}$$
$$= \pi_k^{-1}(f^n(U_k)) \cap \pi_k^{-1}(V_k)$$
$$= \pi_k^{-1}(f^n(U_k) \cap V_k) \neq \varnothing. \quad \square$$

The following theorem now follows easily.

THEOREM 36. *If (X, f) is chaotic, then $(\mathbb{X}(f), \tilde{f})$ is chaotic. Furthermore, $\tilde{f}$ will be a homeomorphism.*

6. Attractors as inverse limits

Attractors play a major role in dynamical systems.

DEFINITION. Let $F:X \to X$ and Λ be a closed subset of X. Then Λ is an *attractor* of F if there exists an open neighborhood N of Λ such that $F(\overline{N}) \subset N$ and $\Lambda = \bigcap_{n \geq 0} F^n(N)$.

A set B is an *invariant set* of F if $F(B) = B$. Note that Λ is an invariant set of F. Intuitively, an attractor is an invariant set to which all nearby orbits converge.

It is often the case that $N = X$. The next theorem represents excellent motivation for using inverse limits in the study of chaos.

THEOREM 37. *If $F:X \to X$ is any one-to-one map, then it is natural to think of $\Lambda = \bigcap_{n \geq 0} F^n(X)$ as the $\varprojlim (X, F)$. That is,*

$$X \xleftarrow{F} X \xleftarrow{F} X \xleftarrow{F} \cdots \Lambda.$$

Explicitly, from Proposition 18 we can think of Λ as the $\varprojlim (F^n, i_n)$; where each i_n is an injection map. Then the diagram

$$
\begin{array}{ccccccccc}
X & \xleftarrow{F} & X & \xleftarrow{F} & X & \xleftarrow{F} & \cdots & & \mathbb{X} \\
\downarrow \text{id} & & \downarrow F & & \downarrow F^2 & & \cdots & & \downarrow F^\infty \\
X & \xleftarrow{i_0} & F(X) & \xleftarrow{i_1} & F^2(X) & \xleftarrow{i_2} & \cdots & & \Lambda
\end{array}
$$

induces a map $F^{\infty}: \mathbb{X} \to \Lambda$ *defined by* $F^{\infty}(x_0, x_1, x_2, \ldots)$ $(= (x_0, F(x_1),$ $F^2(x_2), \ldots) = (x_0, x_0, x_0, \ldots)) = x_0$ *and* F^{∞} *is a homeomorphism.*

We will now study the *solenoid* attractor by first giving its usual presentation in chaos books, and then by giving an inverse limit presentation and demonstrating that it is in fact chaotic.

Let $B^2 = \{(x, y) \in \mathbb{R}^2 : x^2 + y^2 \le 1\}$ and let $D = S^1 \times B^2$. You can think of D as a solid torus imbedded in 3-space. We want to define $F: D \to D$ so that $F(D)$ lies in the interior of D and wraps twice around D. In this case let $S^1 = \mathbb{R} \bmod 2\pi$ so that a point of S^1 equals the angle in radians that the point subtends relative to the positive x-axis. Define $F: D \to D$ by

$$F(\Theta, p) = \left(2\Theta, \frac{p}{10} + \frac{1}{2}e^{i\Theta}\right).$$

Observe that the 2Θ has the effect that $F(D)$ is wrapped twice around D. If $B(\Theta) = \Theta \times B^2$, then $F(B(\Theta))$, $F(B(\Theta + \pi)) \subset B(2\Theta)$. The $\frac{p}{10}$ has the effect that $F(B(\Theta))$ has $\frac{1}{10}$ the diameter as $B(\Theta)$ and the $+\frac{1}{2}e^{i\Theta}$ has the effect that $F(B(\Theta))$ has been translated in $B(2\Theta)$ to the point $\frac{1}{2}$ the distance from the origin to the point $e^{i\Theta}$ which is the point on the boundary of $B(2\Theta)$) which subtends an angle of Θ with the positive x-axis. We clearly have that $F(D) \subset \overset{\circ}{D}$ and that $D \supset F(D) \supset F^2(D) \supset \cdots$.

Let $\Lambda = \bigcap_{n \ge 0} F^n(D)$.

From Theorem 37 we have $D \overset{F}{\leftarrow} D \overset{F}{\leftarrow} D \overset{F}{\leftarrow} \cdots \Lambda$. Recall that $D = S^1 \times B^2$ and define $\pi: D \to S^1$ by $\pi(\Theta, p) = \Theta$, the projection onto the first coordinate. Also recall that $g: S^1 \to S^1$ is defined by $g(x) = 2x$.

LEMMA 38. *The diagram*

$$
\begin{array}{ccccccccc}
D & \overset{F}{\leftarrow} & D & \overset{F}{\leftarrow} & D & \overset{F}{\leftarrow} & \cdots & & \Lambda \\
\downarrow \pi & & \downarrow \pi & & \downarrow \pi & & & & \downarrow \tilde{\pi} \\
S^1 & \overset{g}{\leftarrow} & S^1 & \overset{g}{\leftarrow} & S^1 & \overset{g}{\leftarrow} & \cdots & & \mathbb{S}
\end{array}
$$

commutes and thus induces a map $\tilde{\pi}: \Lambda \to \mathbb{S}$. *Furthermore* $\tilde{\pi}$ *is a homeomorphism. As we have seen before, none of the* π *'s are* 1-1 *but* $\tilde{\pi}$ *is.*

LEMMA 39. *The maps* $\tilde{F}: \Lambda \to \Lambda$ *and* $\tilde{g}: \mathbb{S} \to \mathbb{S}$ *are topologically conjugate by the homeomorphism* $\tilde{\pi}: \Lambda \to \mathbb{S}$.

$$
\begin{array}{ccc}
\Lambda & \overset{\tilde{F}}{\to} & \Lambda \\
\downarrow \tilde{\pi} & & \downarrow \tilde{\pi} \\
\mathbb{S} & \overset{\tilde{g}}{\to} & \mathbb{S}
\end{array}
$$

THEOREM 40. *The map* $(\Lambda, \tilde{F})$ *is chaotic.*

Inverse limits provide a powerful tool for constructing complicated spaces and maps from simple (polyhedral) spaces and maps.

References

1. M. Barge and J. Martin, *Chaos, periodicity, and snakelike continua*, Trans. Amer. Math. Soc. **289** (1985), 355–365.
2. ____, *The construction of global attractors*, Proc. Amer. Math. Soc. (to appear).
3. Morton Brown, *Some applications of an approximation theorem for inverse limits*, Proc. Amer. Math. Soc. **11** (1960), 478–483.
4. R. L. Devaney, *An introduction to chaotic dynamical systems*, Addison-Wesley, Redwood City, Calif., 1987.
5. S. Eilenberg and N. Steenrod, *Foundations of algebraic topology*, Princeton Univ. Press, Princeton, New Jersey, 1952.
6. J. G. Hocking and G. S. Young, *Topology*, Addison-Wesley, Reading, Mass., 1961.
7. R. M. Schori, *Universal snake-like continua*, Proc. Amer. Math. Soc. **16** (1965), 1313–1316.
8. J. van Mills, *Infinite-dimensional topology*, North-Holland, Amsterdam, 1989.
9. R. F. Williams, *One dimensional non-wandering sets*, Topology **6** (1967), 473–487.
10. ____, *Structure of the Lorenz attractors*, Inst. Hautes Études Sci. Publ. Math. **50** (1979), 59–72.

Department of Mathematics, Oregon State University, Corvallis, Oregon 97331

Contemporary Mathematics
Volume **117**, 1991

How Big is the Intersection of Two Thick Cantor Sets?

R. F. WILLIAMS

ABSTRACT. Newhouse introduced the concept of thickness $\tau(C)$ for linear Cantor sets C and proved $C \cap C' \neq \varnothing$ for certain Cantor sets, provided $\tau(C)\tau(C') > 1$. We give examples here where $\tau = \tau(C) = \tau(C')$ and $C \cap C' = \{$one point$\}$, for any τ, $1 < \tau < \sqrt{2}+1$. This bound is sharp as we show that $C \cap C'$ contains a Cantor set provided $\tau(C)$, $\tau(C') > \sqrt{2}+1$. However, our best result with an asymmetric assumption on $\tau(C)$ and $\tau(C')$ probably is not sharp.

Thickness for Cantor sets was introduced by Newhouse in his thesis [**1**] and used by him and others in several important papers [**2, 3, 4**]. It is crucial in dynamical systems in spite of its apparent naiveté. The elementary but deep fact is the

THEOREM (Newhouse [**50**, p. 107]). *If two Cantor sets* C, $C' \subset \mathbb{R}$ *of thickness* τ, τ' *satisfy*

(a) *neither* C *nor* C' *lies in a complementary domain of the other, and*
(b) $\tau \cdot \tau' > 1$,

then $C \cap C' \neq \varnothing$; *in fact, for each* i, $C_i \cap C_i'$ *has a non-empty interior for any defining sequences (see* §2) *of* C *and* C'.

(For $x \in \mathbb{R} - C$, the *complementary domain* of C containing x is the largest open interval containing x and lying in $\mathbb{R} - C$; two of these are unbounded. The end points of these intervals are called *end points* of C; other points of C, *non-end points*. Below we say C and C' are *interleaved* provided condition (a) holds.)

This is an existence theorem and as such it guarantees only *one* point in the intersection. (This suffices for Newhouse's purpose.) But thickness

---

1980 *Mathematics Subject Classification* (1985 *Revision*). Primary 58.
Partially supported by a grant from the National Science Foundation.
This paper is in final form and no version of it will be submitted for publication elsewhere.

seems more analytic than topological, so one expects the correct answer to be some sort of measure of the intersection set. Pondering upon this, the author was surprised that the intersections could indeed be only one point. In this sense, the forced intersection is more topological than analytic, after all. The example found was a bit ad hoc; hence the author returned to this problem to figure out what is going on.

The resolution given here is satisfactory in the symmetric case (i.e., for $\tau = \tau'$), as follows: There is a critical thickness, $\sqrt{2}+1$, such that interleaved Cantor sets of thickness $\tau, \tau' > \sqrt{2} + 1$, intersect in another Cantor set. However, for $\tau, \tau' < \sqrt{2}+1$, $C \cap C'$ need only contain a single point, even for C and C' "regular" or "middle α" Cantor sets.

The critical number arises as a solution of the quadratic equation $\tau^2 - 2\tau = 1$. The resolution in the unsymmetric case (analogous to $\tau\tau' > 1$ in Newhouse's theorem) is less satisfactory. For *regular Cantor sets* one needs $\tau, \tau' \leq \sqrt{2} + 1$ for examples with $C \cap C' =$ one point. In some ways it seems that $\tau\tau' - \tau - \tau' > 1$ would be the correct mixed condition. However, my best result for the intersection to contain a Cantor set uses the following:

$$\begin{cases} \tau\tau' - 2\tau > 1, \\ \tau\tau' - 2\tau' > 1, \end{cases}$$

which for $\tau = \tau'$ reduces to $\tau = \tau' > \sqrt{2} + 1$. Note that it contains some additional information. For example, one can take $\tau = 3$ and $\tau' = 2.34$ and 2.34 is less than $\sqrt{2} + 1$.

CONJECTURE. *Let C and C' be two interleaved Cantor sets of Newhouse thickness τ, τ'. If $\tau\tau' - \tau - \tau' > 1$, then $C \cap C'$ contains a Cantor set.*

Nor do we have examples of interleaved Cantor sets related to the "mixed case"—e.g., where one of C, C' has thickness $> \sqrt{2} + 1$ where $\tau \cdot \tau' > 1$, yet $C \cap C'$ contains only one point.

ADDED IN PROOF. There are 2 preprints answering this completely: Kraft (preprint, University of Cincinnati) and Hunt, Kan, and Yorke (preprint, University of Maryland) have independently found complete solutions of the problem as to which pairs of Cantor sets must have Cantor sets in their intersection.

We have organized the discussion of our examples to help one see what fails in the mixed case. See also our Extremal Proposition.

The paper is organized as follows. We begin with a short section on regular Cantor sets (called middle α-sets in [3]), in which we give our examples. These examples help to motivate the balance of the paper. Our first proof of the basic example is geometric. In §2 we formulate Newhouse thickness in general and state our result. In §3 we state the key ("Cantor") lemma, show how it implies the theorem, and complete the proof; as an addendum we prove the technical "Extremal Proposition."

It is a pleasure to thank Clark Robinson who read an earlier version of this paper and made several helpful suggestions.

1. Regular Cantor sets

For each real number β between 0 and $1/2$ there is a regular Cantor set C defined as follows. Let $\alpha = 1 - 2\beta$, and let C_0 be the unit interval, $[0, 1]$, or a translate of it. Let U_1 be the open middle portion of C_0 of length α so that $C_1 = C_0 - U_1$ is the union of two intervals of length β. Then let U_2 and U_3 be middle portions of these two intervals of the same length $= \alpha\beta$. Then $C_2 = C - (U_2 \cup U_3)$ is the union of four closed intervals each of length β^2. Continue this process indefinitely. See Figure 1.1. Then $C = \bigcap_{i=0}^{\infty} C_i$ is a *regular Cantor set*. The *Newhouse thickness* is β/α; for general Cantor sets, see the formulation in §2 or Newhouse [1, p. 107].

Let E_i be the 2^{i+1} end points of the intervals making up C_i. Then $E = \bigcup_{i=1}^{\infty} E_i$ is the set of C's end points. Clearly the end points of C are countable whereas the non-end points (those in $C - E$) are not.

1.1. EXAMPLE. *Let C be a regular Cantor set with parameter $\beta = \sqrt{2} - 1$, and let C' be the translation of C to the right by β units. Then $C \cap C'$ consists of countably many common end points of C and C' together with a single (non-end) point.*

PROOF. First we identify the special point, $\theta = \beta/(1 - \beta) = \beta + \beta^2 + \beta^3 + \cdots$, and the reflection $S : \mathbb{R} \to \mathbb{R}$ about θ. In coordinates,

$$S(x) = -x + \frac{2\beta}{1 - \beta}.$$

Note that $S(1) = \beta$, so that S maps C_0 to C_0'. It follows that S being an isometry and an involution interchanges the Cantor sets C and C'. Let $\sigma : \mathbb{R} \to \mathbb{R}$ be the contraction $x \mapsto \beta(x - \theta) + \theta$ with fixed point θ. Then σ maps C into a subset of C' and vice versa. To see this, note that

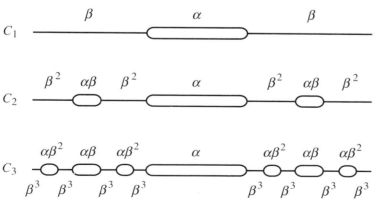

FIGURE 1.1

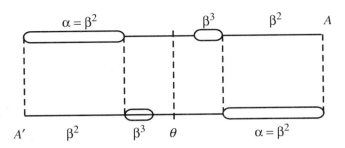

FIGURE 1.2

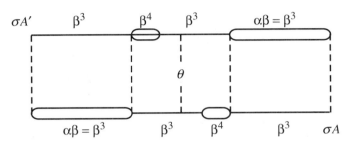

FIGURE 1.3

$\sigma(C_0)$ is the interval $[\beta, 2\beta] \subset C_1'$. Similarly $\sigma(C_0') = \sigma(SC_0) = S\sigma(C_0) = [1 - \beta, 1] \subset C_1$.

The intersection $C_0 \cap C_0' = [\beta, 1]$ whose two end points A and $A' = \theta - A$ are end points of both C and C'. Restricted to this interval our sets are as in Figure 1.2. Note that the similarity σ maps this picture into the following interval (see Figure 1.3) and that $C_1 \cap C_1'$ is this interval together with A and A'.

It follows that

$$
\begin{aligned}
C \cap C' &= [\beta, 1] \cap C \cap C' \\
&= \{A, A'\} \cup \left[\sigma([\beta, 1)) \cap C \cap C' \right] \\
&= \{A, A', \sigma A, \sigma A'\} \cup \left[\sigma^2([\beta, 1)) \cap C \cap C' \right]
\end{aligned}
$$

etc., or that $C \cap C' = \{A, A', \sigma A, \sigma A', \ldots\} \cup \{\theta\}$. □

1.2. EXAMPLE ("Shaving principle"). *There are interleaved Cantor sets* C, C' *of thickness* τ, τ' *for any* $\tau, \tau' < \sqrt{2} + 1$ *such that* $C \cap C'$ *consists of a single point.*

PROOF. We can achieve the single point intersection by removing arbitrarily small neighborhoods of the end points of C and C'. This causes only an arbitrarily small decrease in the thickness.

NOTE. We could allow one of $\tau, \tau' = \sqrt{2} + 1$.

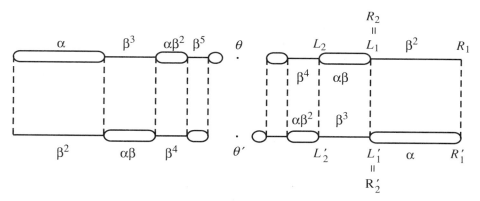

FIGURE 1.4

However, the resulting Cantor sets are not regular (not middle α sets) and next we take the trouble to construct some of these.

1.3. EXAMPLE. *For any $\tau \in (1, \sqrt{2} + 1)$ there are regular interleaved Cantor sets C, C' of thickness τ with $C \cap C' = \{\text{one point}\}$.*

PROOF. We mimic the construction of example (1) with β regarded as a parameter and the sets (and points of interest) as functions of β. Thus our Cantor sets are $C(\beta)$, $C'(\beta)$; their common point $\theta(\beta) = \theta'(\beta)$; and certain end points $R_i(\beta)$, $L_i(\beta)$ of $C(\beta)$ and $R_i'(\beta)$, $L_i'(\beta)$ of $C'(\beta)$. See Figure 1.4.

We will use the point $\theta = \theta'$ as the 0 of our number line and use $R_i = R_i(\beta)$ as points and (positive) numbers, interchangeably. Note next that by symmetry we need only consider the points on the right side of θ. Next we need to show $[L_1, R_1] \subset (L_1', R_1')$, $[L_2', R_2'] \subset (L_2, R_2)$, ... Also that $L_i = R_{i+1}$, $L_i' = R_{i+1}'$, $i = 1, 2, \ldots$ so that it suffices to prove

$$(1.5) \quad \left\{ \begin{array}{l} R_{2i-1}'(\beta) - R_{2i-1}(\beta) > 0 \\ R_{2i}(\beta) - R_{2i}'(\beta) > 0 \end{array} \right\}, \qquad i = 1, 2, \ldots, \ \beta \in (\tfrac{1}{3}, \sqrt{2} - 1).$$

1.6. LEMMA.

$$R_1 = \frac{\beta}{1 + \beta}, \qquad R_1' = \frac{1}{1 + \beta} - \beta,$$

$$R_2 = \frac{\beta}{1 + \beta} - \beta^2 = \beta R_1' \quad and \quad R_2' = \frac{\beta^2}{1 + \beta} = \beta R_1'.$$

PROOF. From the figure,

$$R_1 = \beta^2 + \alpha\beta + \beta^4 + \alpha\beta^3 + \cdots = \beta(\beta + \alpha)(1 + \beta^2 + \beta^4 + \cdots)$$
$$= \frac{\beta(1 - \beta)}{1 - \beta^2} = \frac{\beta}{1 + \beta}.$$

Similarly,

$$R'_1 = \alpha + \beta^3 + \alpha\beta^2 + \beta^5 + \cdots = \frac{\alpha + \beta^3}{1 - \beta^2}$$

$$= \frac{1 - 2\beta + \beta^2 - \beta^2 + \beta^3}{1 - \beta^2} = \frac{1 - \beta}{1 + \beta} - \frac{\beta^2}{1 + \beta}$$

$$= \frac{1}{1 + \beta} - \beta .$$

The remaining computations follow from consideration of the homothety $x \to \beta x$, which sends $C(\beta)$ into a subset of $C'(\beta)$, and $C'(\beta)$ into a subset of $C(\beta)$. $\square$

1.7. COROLLARY. $R_{2i} - R'_{2i} = \beta(R'_{2i-1} - R_{2i-1})$ and $R'_{2i+1} - R_{2i+1} = \beta(R_{2i} - R'_{2i})$, $i = 1, 2, 3, \ldots$.

Thus to finish the verification of Example 3, it suffices to show that $R'_1(\beta) - R_1(\beta) > 0$ for $\beta \in (\frac{1}{3}, \sqrt{2} - 1)$. But this is

$$R'_1(\beta) - R_1(\beta) = \frac{1}{1 + \beta} - \beta - \frac{\beta}{1 + \beta} = \frac{1 - 2\beta - \beta^2}{1 + \beta} > 0. \quad \square$$

We close this section with a computation showing that this construction does not work for $\tau \neq \tau'$, or equivalently, for $\beta \neq \beta'$. We let

$$f(\beta) = \frac{1}{1 + \beta} - \beta \quad \text{and} \quad g(\beta) = \frac{\beta}{1 + \beta}$$

and note the formulas

$$R'_1(\beta') - R_1(\beta) = f(\beta') - g(\beta),$$

$$R_{2i}(\beta) = \beta R'_{2i-1}(\beta), \qquad R'_{2i}(\beta') = \beta' R_{2i-1}(\beta'),$$

and

$$R_{2i+1}(\beta) = \beta R_{2i}(\beta), \qquad R'_{2i+1}(\beta') = \beta' R_{2i}(\beta') .$$

The inequalities corresponding to (1.5) above are then

$$f(\beta') \geq g(\beta),$$

$$\beta f(\beta) \geq \beta' g(\beta'),$$

$$\beta'^2 f(\beta') \geq \beta^2 g(\beta), \ldots .$$

Thus $\beta'^{2n} f(\beta') \geq \beta^{2n} g(\beta)$, $\beta^{2n+1} f(\beta) \geq \beta'^{2n+1} g(\beta')$, and hence

$$\left(\frac{\beta'}{\beta}\right)^{2n} \geq \frac{g(\beta)}{f(\beta')} \quad \text{and} \quad \left(\frac{\beta}{\beta'}\right)^{2n+1} \geq \frac{g(\beta')}{f(\beta)} .$$

It follows that for $\beta \neq \beta'$, $g(\beta) = g(\beta') = 0$, a contradiction.

2. Formulation of the Theorem

2.0. DEFINITIONS. Given a Cantor set $C \subset \mathbb{R}$, we let C_0 be the minimum closed line interval containing C. Then for $x \in C_0 - C$ the complementary interval U_x containing x is the connected component of $C_0 - C$ containing x. There are countably many of these intervals; we suppose they are ordered $U_1, U_2, \ldots$ where $\ell(U_i) \geq \ell(U_{i+1})$. (For the ternary Cantor set, $\ell(U_1) = \frac{1}{3}$, $\ell(U_2) = \ell(U_3) = \frac{1}{9}$, $\ell(U_4) = \cdots = \ell(U_7) = \frac{1}{27}$, etc.)

Let $C_n = C_0 - \bigcup_{i=1}^{n} U_i$ and let B_n, P_n be the interval components of C_n abutting upon U_n. Choose the notation so that $\ell(B_n) \leq \ell(P_n)$; this distinction will be important below. We use Q_n as a neutral name to be one of P_n, B_n. Following (essentially) Newhouse [3, p. 107] we define

$$(2.1) \qquad \tau(c) = \text{thickness of } C = \inf\left\{ \frac{\ell(B_n)}{\ell(U_n)} \;\middle|\; n = 1, 2, \ldots \right\}.$$

LEMMA. *The ambiguity of the choice of the U_i does not affect the definition of Newhouse thickness.*

PROOF. Consider the case that two complementary intervals U_i and U_{i+1} have the same length, say a. It will suffice to show that interchanging the order so that $U_i' = U_{i+1}$ and $U_{i+1}' = U_i$ and $U_j = U_j$ for $j \neq i, i+1$ will not affect the above definitions. This is because the permutation group is generated by transpositions of adjacent elements.

There are two cases. First, in case $(P_{i+1} \cup U_{i+1} \cup B_{i+1}) \cap (P_i \cup U_i \cup B_i) = \varnothing$, there is, a priori, no change whatsoever. In the contrary case, $U_{i+1} \subset P_i \cup U_i \cup B_i = I$ so that U_i and U_{i+1} separate I into three intervals, say of lengths α, β, and γ, where U_i is between, say those of length α and β, and U_{i+1} is between those of length β and γ. Then the thickness ratios occuring for the un-primed system are the same except at stages i and $i+1$. The ratios involved at these stages are α/a or $(\beta+a+\gamma)/a$ and β/a or γ/a whichever is least in both cases. The crucial ratios are among α/a, β/a, and γ/a. For the prime-system, the ratios at the ith and $(i+1)$th stages are

$$(2.2) \qquad \frac{\beta + a + \gamma}{a} \text{ or } \frac{\gamma}{a} \text{ and } \frac{\alpha}{a} \text{ or } \frac{\beta}{a},$$

so that the crucial ratios are among α/a, β/a, and γ/a, again. It follows that the two systems yield the same definition and the invariance of the definition is established.

Newhouse proceeds slightly differently in [3]. His definition (in our language) is

$$(2.3) \qquad \tau_N(C) = \sup_{\mathscr{U}}\left\{ \inf \frac{\ell(B_n)}{\ell(U_n)} \;\middle|\; \mathscr{U} \text{ is an arbitrary ordering}\right.$$

$$\left. \text{of the complementary domains } U_i \text{ of } C \right\}.$$

That this agrees with our definition arises from consideration of the following:

2.4. PROPOSITION. *The supremum in* (2.3) *is attained by* (*each of*) *the orderings given by* (2.0).

PROOF. First, let τ be the infimum given by any ordering as in (2.0). Now assume the proposition is false. It follows that there is an ordering $\mathscr{U}$ giving a larger infimum; in fact there is an ordering $\mathscr{U}$, and a positive integer n such that

(2.4.1)(a) $\ell(U_i) \geq \ell(U_{i+1})$, $U_i, U_{i+1} \in \mathscr{U}$, $i < n$;

  (b) $\ell(U_n) < \ell(U_{n+1})$;

  (c) $\inf\left\{\dfrac{\ell(B_i)}{\ell(U_i)} \,\middle|\, i \leq n + 1 \text{ , using } \mathscr{U}\right\}$

$$> \inf\left\{\frac{\ell(B_i)}{\ell(U_i)} \,\middle|\, \text{ using an ordering as in (2.0), } i \leq n + 1\right\}.$$

Were this not the case, we could reorder $\mathscr{U}$ so that $\ell(U_i) \geq \ell(U_{i+1})$, $i \leq n$ for all n, without changing the infimum in (2.3) for $\mathscr{U}$. But then $\mathscr{U}$ would be one of the orderings in (2.0) with an infimum $> \tau$, which is absurd.

We now consider two cases concerning the U_i, B_i, P_i for the ordering $\mathscr{U}$.

Case (a): $(U_n \cup P_n \cup B_n) \cap (U_{n+1} \cup P_{n+1} \cup B_{n+1}) = \varnothing$. Then the ordering $\{U_i'\}$ in which $U_n' = U_{n+1}$, $U_{n+1}' = U_n$, and $U_i' = U_i$ otherwise, yields exactly the same ratios in (2.3) and thus equality in (2.4.1)(c). This contradiction shows that case (a) cannot hold.

Case (b): $U_{n+1} \subset U_n \cup P_n \cup B_n$. Again defining $\{U_i'\}$ as in case (a) we have the picture in Figure 2.1, where we have labeled the intervals with their lengths, where $a < A$.

But note that

$$\min\left\{\frac{\alpha}{a}, \frac{\beta + A + \gamma}{a}, \frac{\beta}{A}, \frac{\gamma}{A}\right\} = \min\left\{\frac{\alpha}{a}, \frac{\beta}{A}, \frac{\gamma}{A}\right\}$$

$$\leq \min\left\{\frac{\alpha + a + \beta}{A}, \frac{\gamma}{A}, \frac{\alpha}{a}, \frac{\beta}{a}\right\}.$$

But this contradicts (2.4.1)(c), so that our assumption was false, and the proposition is proved.

STANDING HYPOTHESES. C and C' are interleaved Cantor sets in $\mathbb{R}$.

THEOREM I. *Let C and C' be interleaved Cantor sets of thickness τ, τ'. Then*

  (a) *for $\tau, \tau' > \sqrt{2} + 1$, $C \cap C'$ contains a Cantor set;*
  (a') *for $\tau\tau' - 2\tau > 1$ and $\tau\tau' - 2\tau' > 1$, $C \cap C'$ contains a Cantor set;*
  (b) *if $\tau, \tau' \geq \sqrt{2} + 1$ then $C \cap C'$ is infinite; and*
  (c) *there exist examples with*
      (i) *$\tau = \tau' = \sqrt{2} + 1$ and $C \cap C'$ is countable, and*
      (ii) *$C \cap C' = \{\text{a point}\}$, for any τ, τ', $1 < \tau, \tau' \leq \sqrt{2} + 1$, not both $= \sqrt{2} + 1$.*

Recall that (c)(i) is Example 1 of §1 and (c)(ii) is Example 2. Part (b) is less important and can be seen by carefully following the proof of part (a), below. We will say nothing more about part (b).

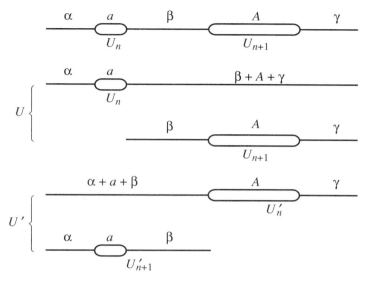

FIGURE 2.1

The next lemma clearly implies (a$'$). As $\sqrt{2}+1$ is a solution to $\tau^2 - 2\tau = 1$, (a$'$) implies (a). In it and in the remainder of the paper, we use the notation introduced at the beginning of this section, C_n, B_n, P_n, Q_n, U_n for C, and C_n', B_n', P_n', Q_n', U_n' for C'.

3. The Cantor Lemma

3.1. CANTOR LEMMA. *If $\tau(C)$, $\tau(C') > \sqrt{2}+1$ and if I is an interval component of $C_{m-1} \cap C_{n-1}'$, then for some larger intergers, α, β, $C_\alpha \cap C_\beta' \cap I$ has (at least) two interval components.*

PROOF. We proceed indirectly. Fix such an I for which there are no such α and β. Note that we could have I an interval component of $C_{m-1} \cap C_{n-1}'$ for several values of m and n. But not for infinitely many, since then one of the two Cantor sets would contain I, which is absurd. Thus we may assume m and n are both maximal. Now $I \subset B_m \cup U_m \cup P_m$ since $C_m = C_{m-1} - U_m$. By our assumption $I \not\subset C_m$ so that $I \not\subset B_m \cup P_m$. Hence one of the two intervals B_m, P_m, say Q_m, contains $I - U_m$. Similarly, Q_n' is chosen to be one of the B_n', P_n' so that $Q_n' \supset I - U_n'$.

Now $\overline{U}_m$ and $\overline{U}_n'$ both contain an end point of I. These could not be the same, as then I would intersect either both of B_m and P_m or both of B_n' and P_n'. (See Figure 3.2.) Hence our picture is as shown in Figure 3.3. The orientation of this figure could be reversed, but that is unimportant. From it we easily see that $\overline{U}_n'$ contains an end point of Q_m, as otherwise int(I) would extend beyond $\overline{U}_n'$ to intersect both B_n' and P_n'. Similarly $\overline{U}_m$ contains an end point of Q_n'. Note then that the locus of possible end points for I are $\overline{U}_m$ and $\overline{U}_n'$, respectively.

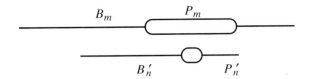

FIGURE 3.2

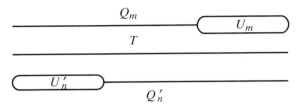

FIGURE 3.3

Say U_r is the first complementary interval of C hitting Q_m, and U_s' the first one of C' hitting Q_n'. Now consider the four intervals B_r, P_r, B_s', and P_s'. All four intersect I by maximality of m and n, but by our assumption the intersection $(B_r \cup P_r) \cap (B_s' \cup P_s')$ contains one and only one nondegenerate interval, say I_1. This leads us to two (the choice of names will seem more appropriate later on) dichotomies:

First,
$$B: P_r \subset \overline{U}_n' \cup \overline{U}_s' \quad \text{or} \quad P: B_r \subset \overline{U}_n' \cup \overline{U}_s'.$$

Secondly,
$$B': P_s' \subset \overline{U}_m \cup \overline{U}_r \quad \text{or} \quad P': B_s' \subset \overline{U}_m \cup \overline{U}_r.$$

For these various intervals we will use lower case letters to denote their lengths—e.g., $\ell(B_r) = b_r$. We also introduce the "Newhouse thinness" parameters, $\beta = 1/\tau(C)$ and $\beta' = 1/\tau(C')$.

To prove the Cantor Lemma we use the conditions $\beta\beta' + 2\beta < 1$ and $\beta\beta' + 2\beta' < 1$, which are equivalent to $\tau\tau' - 2\tau > 1$ and $\tau\tau' - 2\tau' > 1$. Note the following relations between the τ's:

REMARK. $(\tau\tau' - \tau - \tau')^2 - (\tau - \tau')^2 = (\tau\tau' - 2\tau)(\tau\tau' - 2\tau') = (\tau^2 - 2\tau)(\tau'^2 - 2\tau')$.

Thus several cases occur; unfortunately we will have to consider unbounded sequences of cases. Thus we introduce notation as follows: first our cases come with subscripts, e.g., BP_i', PB_i', or PP_i', $i = 0, 1, 2, \ldots$. Next, let r_1, s_1 be the r, s of the previous paragraphs and introduce $r_2 = t$ so that U_t is the first complementary interval of C to hit I_1, $s_2 = v$ so that U_v' is the first complementary interval of C' to hit I_1. This leads to I_2, r_3, s_3, etc. We will avoid double subscripts by simply suppressing the middle guy. Thus we have complementary intervals $U_1, U_2, \ldots, U_1', U_2', \ldots$, and component intervals $P_1, P_2, \ldots, B_1, B_2, \ldots, P_1', P_2', \ldots, B_1', B_2', \ldots$. Finally

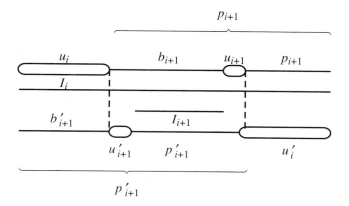

FIGURE 3.5

we denote the lengths of these intervals by the corresponding lower case letter.

3.4. LEMMA. *Assume* $\beta\beta' + 2\beta$ *and* $\beta\beta' + 2\beta' \leq 1 - \varepsilon$. *Then*

$$q_{i+1} + q'_{i+1} \geq q_i + q'_i - (b_i + b'_i) + (\beta' b'_i + \beta b_i) + \varepsilon(b_i + b'_i) .$$

PROOF. Assume case B_{i+1}. Then

$$
\begin{aligned}
q_{i+1} = b_{i+1} = q_i &- p_{i+1} - u_{i+1} \\
&\geq q_i - p_{i+1} - \beta p_{i+1} , \quad \text{by the Newhouse condition} \\
&\geq q_i - u'_i - \beta u'_i , \quad \text{by case } B_{i+1} \text{ (see Figure 3.5)} \\
&\geq q_i - \beta' b'_i - \beta\beta' b'_i , \quad \text{by Newhouse again} \\
&\geq q_i - (2\beta' + \beta\beta')b'_i + \beta' b'_i \\
&\geq q_i - (1 - \varepsilon)b'_i + \beta' b'_i \\
&\geq q_i - b'_i + \beta' b'_i + \varepsilon b'_i .
\end{aligned}
$$

In case P_{i+1},

$$
\begin{aligned}
q_{i+1} = p_{i+1} = q_i &- b_{i+1} - u_{i+1} \\
&\geq q_i - b_{i+1} - \beta b_{i+1} \\
&\geq q_i - u'_i - \beta u'_i \\
&\geq q_i - \beta' b'_i - \beta\beta' b'_i \\
&\geq q_i - (2\beta' b'_i + \beta\beta')b'_i + \beta' b'_i \\
&\geq q_i - b'_i + \beta' b'_i + \varepsilon b'_i .
\end{aligned}
$$

Similarly, for B'_{i+1} and P'_{i+1}, we conclude $q'_{i+1} \geq q'_i - b_i + \beta b_i + \varepsilon b_i$. Adding these two inequalities, we obtain the desired inequality.

3.6. LEMMA. $b_{i+1} \leq \beta' b'_i$ *and* $b'_{i+1} \leq \beta b_i$.

PROOF. In any case above we saw that $b_{i+1} \leq u_i' \leq \beta' b_i'$, which proves the first inequality. The second follows by symmetry.

3.7. LEMMA. *For* n, m *integers*, $n > m$,

$$q_{n+1} - q_{n+1}' \geq (q_m - b_m) + (q_m' - b_m') + \varepsilon \sum_{i=m}^{n} (b_i + b_i') \ .$$

PROOF. Summing the inequalities of Lemma 3.4 from m to n, we get

$$\sum_{i=m}^{n} (q_{i+1} + q_{i+1}') \geq \sum_{i=m}^{n} (q_i + q_i') - \sum_{i=m}^{n} [(b_i + b_i') - (\beta' b_i' + \beta b_i)]$$
$$+ \varepsilon \sum_{i=m}^{n} (b_i + b_i') \ .$$

Cancelling we obtain

$$q_{n+1} + q_{n+1}' \geq q_m + q_m' - (b_m + b_m') + \sum_{i=m}^{n-1} [(\beta' b_i' - b_{i+1}') + (\beta b_i - b_{i+1})]$$
$$+ (\beta' b_m' + \beta b_m) + \varepsilon \sum_{i=m}^{n} (b_i + b_i')$$

$$\geq (q_m - b_m) + (q_m' - b_m') + \sum_{i=m}^{n-1} [(\beta' b_i' - b_{i+1}') + (\beta b_i - b_{i+1})]$$
$$+ (\beta' b_m' + \beta b_m) + \varepsilon \sum_{i=m}^{n} (b_i + b_i')$$

$$> (q_m - b_m) + (q_m' - b_m') + \varepsilon \sum_{i=m}^{n} (b_i + b_i') \ .$$

This proves Lemma 3.7.

PROOF OF 3.1, THE CANTOR LEMMA. We have shown that the assumption that the Cantor Lemma is false leads to the inequality of Lemma 3.7. But this means that the intervals $I_m \supset I_{m+1} \supset I_{m+2} \supset \cdots$ have lengths bounded below by $\varepsilon(b_m + b_m' + \cdots + b_n + b_n')$ which contradicts the fact that C (as well as C') is a Cantor set. This completes the proof.

Note that we have proved more:

EXTREMAL PROPOSITION. *If* C *and* C' *are interleaved Cantor sets such that*

(1) $C \cap C'$ *contains no Cantor set*,
(2) *their thicknesses satisfy* $\tau\tau' - 2\tau \geq 1$ *and* $\tau\tau' - 2\tau' \geq 1$,
then $\tau = \tau' = \sqrt{2} + 1$.

PROOF. The left-hand side of the inequality of the lemma must go to zero as $n \to \infty$. Thus the right must go to zero, so that $q_m = b_m$ and $q_m' = b_m'$

for $m \geq 0$, and $\varepsilon = 0$. That $\varepsilon = 0$ tells us that $\tau\tau' - 2\tau = \tau\tau' - 2\tau' = 1$. Hence $\tau = \tau'$ and $\tau = \tau' = \sqrt{2} + 1$, and the proof is complete.

No way of exploiting the facts that $q_m = b_m$ and $q'_m = b'_m$ is known. These would tend to say that the original Cantor sets are "almost regular."

REFERENCES

1. S. Newhouse, *Nondensity of axiom $A(a)$ on S^2*, Proc. Sympos. Pure Math., vol. 14, Amer. Math. Soc., Providence, R. I., 1970, pp. 191–203.
2. ____, *Diffeomorphisms with infinitely many sinks*, Topology **13** (1974), 9–18.
3. ____, *The abundance of wild hyperbolic sets and non-smooth stable sets for diffeomorphisms*, Inst. Hautes Études Sci. Publ. Math. **50** (1979), 101–151.
4. C. Robinson, *Bifurcation to infinitely many sinks*, Comm. Math. Phys. **90** (1983), 433–459.

DEPARTMENT OF MATHEMATICS, THE UNIVERSITY OF TEXAS AT AUSTIN, AUSTIN, TEXAS 78712

Contemporary Mathematics
Volume 117, 1991

Problems in Dynamics on Continua

MARCY BARGE AND MORTON BROWN

Most of these problems were raised at the Arcata meeting in June 1989. Some are much older. The last set of problems on the pseudoarc was compiled by Wayne Lewis.

1. **Inverse limits and interval maps.** (f denotes a map of the compact interval I to itself, (I, f) is the associated inverse limit space, and $\hat{f} \colon (I, f) \to (I, f)$ is the shift homeomorphism.)

1.1 (J. Martin) Suppose that $M \in I$ is an infinite minimal set for f and that no proper invariant subinterval of I contains M. Let $\hat{M} = \{(m_0, m_1, \dots) \in (I, f) \colon m_i \in M \text{ for all } i\}$. Is each point of $\hat{M}$ an endpoint of (I, f)?

1.2 (J. Martin) For what f is it true that for each $\varepsilon > 0$, f is conjugate to a map that is within ε of the identity?

1.3 (J. Martin) If C is the space of maps of I onto itself, DEC is the set of f for which (I, f) is decomposable, and INDEC is the set of f for which (I, f) is indecomposable:
   (a) Is DEC the closure of its interior?
   (b) Is INDEC the closure of its interior?

1.4 (J. Ding) Let $S(x, m) = 4mx(1 - x)$ and $T(x, b) = \{2bx$ if $0 \le x < 1/2$, $2b(1 - x)$ if $1/2 < x \le 1\}$. For what m, b, $0 \le m$, $b \le 1$, are $S(\ , m)$ and $T(\ , b)$ conjugate maps of I?

1.5 (M. Barge) For what f is it the case that (I, f) can be embedded in the plane so that $\hat{f}$ extends to a diffeomorphism?

1.6 (W. Transue) Given $f \colon I \to I$ and $\varepsilon > 0$, is there a $g \colon I \to I$ within ε of f such that all subcontinua of (I, g) of diameter less than ε are pseudo-arcs?

1.7 (S. Baldwin) Let $S = \{1, 2, \dots, n\}$, let p, q map S into S, and let $f(p)$ and $f(q)$ map $[1, n]$ into itself by linear extension. Find an effective algorithm which decides when the induced inverse limit spaces are homeomorphic.

©1991 American Mathematical Society
0271-4132/91 $1.00 + $.25 per page

2. Homeomorphisms of continua.

2.1 (R. F. Williams) Does there exist a chainable continuum admitting an expansive homeomorphism? What if the continuum is planar?

2.2 (M. Barge) Does every homeomorphism of an hereditarily decomposable chainable continuum have zero topological entropy?

2.3 (J. Mayer) What continua in E^2 (E^3) and what homeomorphisms of them admit extension to E^2 (E^3) in such a way that the given continuum is an attractor?

2.4 (K. Kuperberg) Does there exist a homogeneous locally compact metric space (continuum) X such that if $p \neq q \in X$ then there is no homeomorphism $h: X \to X$ with $h(p) = q$ and $h(q) = p$?

3. Homeomorphisms of the Euclidean space.

3.1 (J. Martin) Suppose D is a 2-disk and h is a homeomorphism of D into itself. Find necessary and/or sufficient conditions that the intersection of forward iterates of h is indecomposable.

3.2 (M. Barge) Let f be a diffeomorphism of the plane with fixed hyperbolic saddle p, and suppose that one branch of the unstable manifold of p contains a nontopologically transverse homoclinic point. Is it necessarily the case that the closure of that branch of the unstable manifold is not chainable?

3.3 (E. Slaminka) Let h be an orientation-preserving homeomorphism of the plane with isolated fixed point p of fixed point index $-n < 1$. Let D be a disk containing p in its interior and containing no other fixed points. Do there exist $n + 1$ continua S_i and $n + 1$ continua U_i such that:
  (a) S_i and U_i are contained in D;
  (b) $p \in S_i, U_i$;
  (c) $S_i \cap S_j = \{p\} = U_i \cap U_j$ for $i \neq j$:
  (d) none of the S_i or U_i separate the plane;
  (e) if $x \in S_i$ then x goes to p under iteration of h, and if $x \in U_i$ then x goes to p under iteration of h^{-1}; and
  (f) $h(S_i) \subset S_i$ and $h^{-1}(U_i) \subset U_i$?

3.4 (L. Oversteegen) Let f be a recurrent homeomorphism of the 2-disk D. If M is a minimal invariant subcontinuum of D, is M homeomorphic to a circle (or a point) and f restricted to M conjugate to a rotation?

3.5 (L. Oversteegen, modification of a problem by Birkhoff) Let f be an analytic measure-preserving homeomorphism of the 2-torus. Is f conjugate to a rotation?

3.6 (L. Oversteegen) Let f be a homeomorphism of the open set U of the plane onto itself and let X be the set of all points of U whose forward orbit is dense in U. Suppose that X is nonempty.
  (a) Does X contain a locally connected continuum?
  (b) Does X contain a separating continuum?

3.7 (Aarts) Let X be a one-to-one continuous image of $\mathbb{R}$ and suppose that X is homogeneous and nonlocally connected. Is there a neighborhood of each point of X homeomorphic with the product of S with $\mathbb{R}$ for some closed subset S of X?

3.8 (Aarts) Let Π be a flow on $\mathbb{R}^n$. Does each point of $\mathbb{R}^n$ have a flowbox neighborhood homeomorphic with the n-dimensional disk? (Yes, if $n \leq 3$; and there is a flow on $\mathbb{R}^4$ so that at some points there is no local section homeomorphic with the 3-disk.)

3.9 (P. Blanchard) Let $P(z, \lambda) = \lambda z + z^2$ and assume that $\lambda = \exp(2\pi i\alpha)$, α well approximated by rationals. Then the Julia set $J(P(\ ,\lambda))$ is not locally connected.
  (a) What is the topological structure of $J(P(\ ,\lambda))$?
  (b) What are the dynamical properties of $P(\ ,\lambda)$ restricted to $J(P(\ ,\lambda))$?

3.10 (M. Brown) Suppose a homeomorphism of the plane has the origin both as a fixed point and an isolated periodic point. Is there a Jordan domain containing the origin whose intersection with its image is connected?

3.11 (M. Brown) Is the space of fixed point free orientation-preserving homeomorphisms of the plane path connected? Locally connected? Connected?

3.12 If h is an orientation-preserving homeomorphism of the plane with no fixed points, what are the conditions under which two points of the plane lie in a continuum disjoint from its image?

3.13 (M. Herman) Is there a homeomorphism of Euclidean n-space ($n > 2$) with every orbit dense? Is there a minimal flow on the 3-sphere?

3.14 (M. Herman) Is there a homeomorphism of the plane minus a point with every orbit dense?

3.15 (P. Boyland, J. Franks) Let p be a periodic point of an orientation-preserving homeomorphism of the plane. Is there always a fixed point x such that the "rotation number" of p around x is *not* equal to 1?

3.16 (M. Brown) Let h be an orientation-preserving homeomorphism of the plane, let p be an isolated fixed point of index greater than 1, and let U be a neighborhood of p. Is there always a point different from p whose total orbit lies in U?

3.17 (M. Barge) Suppose that h is an orientation-preserving homeomorphism of the plane, that M is a continuum invariant under h having precisely two complementary domains, and that there is a prime end associated with M that is periodic of least period n under the action on prime ends induced by h. Is there necessarily a periodic point of least period n in M?

3.18 (D. Mauldin) Let T map the annulus $S^1 \times \mathbb{R}$ to itself by

$$T(\exp(2\pi ix), y) = (\exp(2\pi iAx, B(y - \phi(x))))$$

where $A > 1$ is an integer, $0 < B$, and $\phi\colon \mathbb{R} \to \mathbb{R}$ is continuous with period 1 (say $\phi(x) = \mathrm{dist}(x, \mathbb{Z})$ or $\phi(x) = \cos 2\pi x$). If $B < 1$, is the intersection of the forward iterates of $S^1 \times [-B/(1 - B), B/(1 - B)]$ an indecomposable continuum?

3.19 (E. Burgess) Suppose that a cofrontier is the union of a continuous collection of homogeneous tree-like continua that are mutually homeomorphic. Must the elements of the collection be pseudo-arcs?

3.20 (R. Walker) Let F be a cofrontier invariant under the orientation-preserving homeomorphism h of the plane. Let G, H be the induced maps on the circles of prime ends associated with F. If both G and H are conjugate to irrational rotation, must F be a minimal set? Can G and H be conjugate to irrational rotations but with different rotation numbers?

3.21 (Brechner, et al.) If G and H (as in (3.20)) are conjugate to the same irrational rotation, is h recurrent on F?

4. Other problems.

4.1 A homeomorphism h of a compact space X is *standard* if there are fixed points $A(h)$ and $R(h)$, not necessarily distinct, such that h^n (resp. h^{-n}) converges uniformly to the constant $A(h)$ (resp. $R(h)$) on compact subsets of $X - \{R(h)\}$ (resp. $X - \{A(h)\}$) as $h \to \infty$. Is there a group G of homeomorphisms of the pseudo-arc P such that:

   (a) every $g \in G - \{\mathrm{id}\}$ is standard, and

   (b) for every $x, y \in P$ and every $\varepsilon > 0$, there is a $g \in G$ such that $A(g)$ is ε-close to x and $R(g)$ is ε-close to y?

4.2 (R. F. Williams) Let $L(n, m)$ be the "knot holder" with semi-flow φ_t defined by Figure 1, in which some orbits flow out of the bottom and are lost and there are n twists on the left and m twists on the right ($\underset{\frown}{\cup}$ is positive, $\bigwedge$ is negative). It is known that the knots (closed orbits) in $L(0, m)$ are prime for $m \geq 0$, that $L(n, m)$ contains nonprime knots for $n \neq 0 \neq m$, and that $L(0, -1)$ contains composite knots.

   (a) Does $L(0, -2)$ contain composite knots?

   (b) Does any $L(n, m)$ contain knots which are composites of more than two knots?

4.3 (R. F. Williams) Two Lorenz attractors A_X and A_Y are not homeomorphic if:

   (a) the kneading sequences of X are different from those of Y, and

   (b) there is no homeomorphism $h\colon A_X \to A_Y$ which does not induce the identity on the fundamental group.

Is (b) necessary?

4.4 (Oversteegen) Let (X, d) be an arcwise connected, locally arcwise connected metric space. Under what additional hypotheses will there

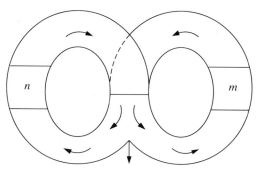

FIGURE 1

exist a compatible convex metric? (Known: X compact (Bing), X uniquely arcwise connected (Mayer and Oversteegen).)

4.5 (Ancel) Let X be a Peano continuum and let $1 \le n \le \infty$. An embedding $e: X \to [0, 1]^n$ is *rectified* if nearby points in $e(X)$ are connected by short (in the sense of arclength) rectifiable arcs in $e(X)$. For $1 \le n \le \infty$, is it true that every embedding of a Peano continuum into $[0, 1]^n$ can be approximated by rectifiable embeddings?

4.6 (Bellamy) Does every deformation of a tree-like continuum have a fixed point?

4.7 (Hagopian) Let M be a hereditarily decomposable continuum that does not contain a simple closed curve. Does every deformation of M have a fixed point?

4.8 (W. Transue) Given that X is a tree-like plane continuum is there an $\varepsilon > 0$ so that if f is a map of X into X with $d(x, f(x)) < \varepsilon$ for all x, then f has a fixed point?

4.9 (Manka)
  (a) Does an arcwise connected plane continuum have the fixed point property if and only if it is simply connected?
  Does the product $X \times Y$ have the fixed point property if:
  (b) X and Y are one-dimensional arcwise connected continua with the fixed point property?
  (c) X and Y are λ-dendroids?

4.10 (W. T. Ingram)
  (a) If M is a tree-like continuum and f is a map of M into M, does f have a periodic point?
  (b) If M is a nonseparating plane continuum and f is a map of M into M, does f have a periodic point?

4.11 (V. Akis) Let Y be the simple triod with a spiral. Does the cone over Y have the fixed point property?

4.12 (V. Akis and C. Pugh) Let H denote the collection of all homeomorphisms of the Cantor set C onto itself. For each $h \in H$ consider the equivalence relation $\sim_h$ on $C \times [0, 1]$ defined by $(x, 1) \sim_h (h(x), 0)$

for all $x \in C$. Characterize all continua of the type $C \times [0, 1]/ \sim_k$ for $h \in H$.

4.13 (Gutek) Let $h: C \to C$ be a homeomorphism of the Cantor set C and, for $Y \subset C$, let $P(Y, h)$ be the statement: for every two nonempty separated subsets Y_1 and Y_2 of Y such that $Y_1 \cup Y_2 = Y$, either $h(Y_1) \cap Y_2 \neq \varnothing$ or $h(Y_2) \cap Y_1 \neq \varnothing$. Suppose that $P(C, h)$ is true but $P(Y, h)$ is false for every proper closed subset Y of C with $\mathrm{int}(Y) \neq \varnothing$. Is there a point in C with a dense orbit?

4.14 What planar continua admit minimal homeomorphisms?